MOLECULAR BASIS OF BIOLOGICAL DEGRADATIVE PROCESSES

Academic Press Rapid Manuscript Reproduction

Proceedings of the Symposium
held at The University of Connecticut
Health Science Center
May 1977

MOLECULAR BASIS OF BIOLOGICAL DEGRADATIVE PROCESSES

EDITED BY

Richard D. Berlin

Department of Physiology
The University of Connecticut Health Center
Farmington, Connecticut

Heinz Herrmann

Department of Biology
The University of Connecticut
Storrs, Connecticut

Irwin H. Lepow

Department of Medicine
The University of Connecticut Health Center
Farmington, Connecticut

Jason M. Tanzer

Department of Restorative Dentistry
The University of Connecticut Health Center
Farmington, Connecticut

ACADEMIC PRESS   New York   San Francisco   London   1978
A Subsidiary of Harcourt Brace Jovanovich, Publishers

ACADEMIC PRESS, INC.
111 Fifth Avenue, New York, New York 10003

United Kingdom Edition published by
ACADEMIC PRESS, INC. (LONDON) LTD.
24/28 Oval Road, London NW1   7DX

Library of Congress Cataloging in Publication Data
Main entry under title:

Molecular basis of biological degradative processes.

Based on papers presented at a symposium held at
the University of Connecticut Health Center,
Farmington, May 19-20, 1977.
    1.   Proteolytic enzymes—Congresses.
2. Lysosomes—Congresses.   3.   Dental caries—
Congresses.   4.   Bone resorption—Congresses.
I.   Berlin, Richard D.   [DNLM: 1.   Biodegradation
—Congresses.   2.   Molecular biology—Congresses.
QH506 M7186 1977]
QP609.P78M64   574.1'33   78-21892
ISBN 0-12-092150-2

PRINTED IN THE UNITED STATES OF AMERICA

78 79 80 81 82   9 8 7 6 5 4 3 2 1

CONTENTS

CONTRIBUTORS

Numbers in parentheses indicate the pages on which the authors' contributions begin.

K. FRANK AUSTEN (125), Harvard Medical School and Robert B. Brigham Hospital, Boston, Massachusetts

WILLIAM H. BOWEN (261), Caries Prevention and Research Branch, National Caries Program, National Institute of Dental Research, National Institutes of Health, Bethesda, Maryland

ROSCOE O. BRADY (39), National Institute of Neurological and Communicative Disorders and Stroke, National Institutes of Health, Bethesda, Maryland

EARL W. DAVIE (115), Department of Biochemistry, University of Washington, School of Medicine, Seattle, Washington

CHRISTIAN de DUVE (25), The Rockefeller University, New York, and International Institute of Cellular and Molecular Pathology, Brussels

R. M. FRANK (277), Groupe de Recherches Inserm U 157, Faculté de Chirurge Dentaire, Université Louis Pasteur, Strasbourg

KAZUO FUJIKAWA (115), Department of Biochemistry, University of Washington, School of Medicine, Seattle, Washington

PAUL GOLDHABER (313), Harvard School of Dental Medicine, Peter Brent Brigham Hospital, Boston, Massachusetts

RONALD L. HEIMARK (115), Department of Biochemistry, University of Washington, School of Medicine, Seattle, Washington

LUCILLE S. HURLEY (1), Department of Nutrition, University of California, Davis, California

AARON JANOFF (225), Department of Pathophysiology, Health Sciences Center, State University of New York at Stony Brook, Stony Brook, New York

WALTER KISIEL (115), Department of Biochemistry, University of Washington, School of Medicine, Seattle, Washington

BRUCE D. KORANT (171), E. I. du Pont de NeMours and Company

KOTOKU KURACHI (115), Department of Biochemistry, University of Washington, School of Medicine, Seattle, Washington

HANS J. MÜLLER-EBERHARD (65), Department of Molecular Immunology, Research Institute of Scripps Clinic, La Jolla, California

LUKA RABADJIJA (313), Harvard School of Dental Medicine, Peter Brent Brigham Hospital, Boston, Massachusetts

E. REICH (155), The Rockefeller University, New York

GEORGE SZABÓ (313), Harvard School of Dental Medicine, Peter Brent Brigham Hospital, Boston, Massachusetts

J. C. VOEGEL (277), Groupe de Recherches Inserm U 157, Faculté de Chirurgie Dentaire, Université Louis Pasteur, Strasbourg

ROGER W. YURT (125), Harvard Medical School and Robert B. Brigham Hospital, Boston, Massachusetts

PREFACE

This book is based on a symposium given at the University of Connecticut Health Center in May 1977. The contributions have been extended and in some cases updated for final publication.

The symposium celebrated 10 years of development at the Health Center. The topic "Molecular Basis of Biological Degradative Processes," includes major areas that have become a fabric underlying much of both basic and applied research in human biology.

The concept of "catabolism" is an ancient one, of course, and may be traced to the Atomists, Democritus, and Lucretius, and through the nineteenth century in the study of fermentation. Applied to animal structure the traditional emphasis was on decay or degradation. In modern terms catabolism was given new meaning, first, in the context of "steady-state turnover of proteins" in which it was seen to balance or control synthesis; and second, in embryological studies in which programmed cell death and degradation was seen as an ordered and essential mechanism of tissue remodeling. Some aspects of the latter process and the influence of nutrition are given in the chapter by Professor Lucille Hurley. The relationship of the lysosome to such remodeling and the broader significance of this organelle for the processing of intra- and extracellular material is described by Professor de Duve. Dr. Roscoe Brady describes the specific application of the lysosome concept to the description of deficiency diseases of lipid catabolism.

A major part of current understanding of "degradative" processes has come from an ever-deepening study of proteolysis. The activation of intestinal proteases by cleavage and release of peptides (e.g., trypsinogen to trypsin) has been known for decades. More recently, the coupling of a series of proteolytic steps into a cascade providing for the amplification of an initial event has been recognized as fundamental in the response of the organism to immune or mechanical stimuli. Examples of cascade phenomena are discussed in detail by Professor Müller-Eberhard (complement), Davie, (coagulation and fibrinolysis), and Austen (activation and function of mast cells). In addition, proteases with high specificities and unique biological functions have been discovered and other well-known enzymes endowed with new significance. Here, Professor Reich discusses the role of plasminogen activator in normal reproductive biology. Added to these secreted proteases and those of the lysosome has come recognition of a set of highly specific intracellular proteases that are involved in the processing of proteins for secretion or assembly. Dr. Bruce Korant reviews the considerable evidence for the essential action of specific endoproteases in the production of animal viruses.

Finally, emphasis is given to degradative processes as they operate on the extracellular matrix. A pathologic role is suggested for secreted neutrophil elastase in the breakdown of the elastin of the lung in the model system described by Professor Janoff. The concerted action of bacterial metabolism and the mode of attack on tooth structure to produce dental caries are analyzed by Drs. Bowen and Frank; and the degradation or remodeling of bone, by Professor Goldhaber.

The discussion of these diverse topics by leading figures in their development provides a unique overview of basic and applied research in "catabolic" processes as they have now come to be understood.

MOLECULAR BASIS OF BIOLOGICAL DEGRADATIVE PROCESSES

DEVELOPING ORGANISMS AS MODEL SYSTEMS FOR THE STUDY OF DEGRADATIVE PROCESSES

Lucille S. Hurley

University of California

Degradative processes are important in normal morphogenesis, both as the degradation that is part of turnover, and as massive cell death. Examples of massive programmed cell death in the modelling of structure are presented, notably the development of the limbs, the palate and the aortic arches. Abnormalities of cell death and tissue degeneration, either excessive or insufficient in quantity, or incorrect in timing, result in development of anomalous structures. Such examples are seen in various mutants. The mechanisms underlying cell death and its control are discussed, and possible experimental models for its study are presented. Attention is focused on nutrient deficiencies during prenatal development and research is discussed on congenital abnormalities resulting from deficiencies of zinc, magnesium, and manganese, and of mutant gene-trace element interactions involving manganese and copper.

A. "In the Midst of Life We are in Death" (1)

To link degradative processes and development seems almost a contradiction in terms. Although development is not synonymous with growth, the developing organism undergoes a remarkable enlargement of total body mass,

especially in prenatal life. The increase in both cell
number and cell size involved in the development from
a single fertilized ovum to the newborn animal is well
known. How then, can one think of degradative processes in the face of such an obvious drive toward synthesis?

One aspect of degradative processes which comes
immediately to mind is the catabolism that is a normal
part of turnover even in an actively synthesizing
system. Such degradation plays a role in the biochemical reactions which underlie development, and control
of its rate in relation to synthesis may be important
in regulating the differential growth of cells, tissues,
and organs.

There is another type of degradative process in
development, however, which is more dramatic. This is
massive cell death and tissue degeneration, an important
phenomenon in the normal development of structure. In
the words of Elizabeth Barrett Browning, "...Life is
perfected by Death" (2).

In my discussion today, I am going to outline some
manifestations of the role of cell death and degradation
in normal morphogenesis. I will then present a few
examples of the consequences to the developing organism
of abnormal degenerative processes. Finally, I would
like to present some examples from my own research of
approaches that may be useful as experimental models for
the study of degradative processes.

I. CELL DEATH IN NORMAL MORPHOGENESIS

Cell death is an important phenomenon in the normal
development of structure during embryonic and fetal
development. Glücksmann, in a review published in 1951
(3), listed 74 different sites in which embryonic cell
death could be localized. These included such examples
as the presumptive neural tissue and primitive node of
the early ambryo, the neural plates, neural groove and
neural tube, various portions of the eye, ear, nose,
branchial arches, and intestinal tract. Formation of
the larynx, trachea, and lung also involves degeneration
and cell death. The notochord which gives rise to the

skeleton, as well as the somites which give rise to the future muscle system, also undergo degeneration involving cell death of portions of the tissue. And, bone itself, of course, during postnatal as well as prenatal life, involves degradative processes as the growing bone is remodeled in order to maintain its proper proportions (4).

Cell death and tissue degeneration are involved in the union or the separation of parts of the body or of rudiments. Thus the union of the body halves appear to involve cell death in the midventral line, while cell death in the olfactory tube and palate and in the notochord are part of the process of the detachment of these structures from their previous connections. Similarly, degeneration of part of the aorta seems to be necessary for its bifurcation. Another important aspect of morphogenetic degeneration is in the formation of lumina in solid or partly occluded organs. Many of the tubular structures in the body develop first as solid organs and only later is the lumen formed. Thus, the salivary and esophageal glands, the duodenum, colon, and vagina, as well as vascular rudiments, show cell death which appears to be related to the formation of the lumen. In general, both invaginations and evaginations occurring during embryonic morphogenesis involve tissue degeneration.

The regression of transient structures is also important. Transient structures are those which are formed but which subsequently regress completely, as, for instance, the tail of the tadpole, or, in vertebrates, the mesonephros, which is present at one stage of the development of the urinary system. Thus, regression of various structures such as the pharyngeal, branchial, and cloacal membranes is associated with cell death. Similarly, epidermal seams of various structures such as lids, lips, and external auditory meatus regress, apparently through cell death, as do also vestigial organs. The differentiation of various tissues and organs per se, also involves degeneration. For example, the differentiation of male sex organs requires the regression of the Mullerian duct, while the development of female sex organs requires the degeneration of the Wolffian duct (3,5,6).

B. Limb Development

One of the systems in which the role of cell death in morphogenesis has been best studied in the limb. In the chick embryo, Saunders and Fallon (7) have shown that in the limb bud and leg primordia of the chick these is massive cell death in the interdigital spaces which is involved in the formation of the digits (8). In the duck, however, the distribution of the necrosis is quite different (see Fig. 1).

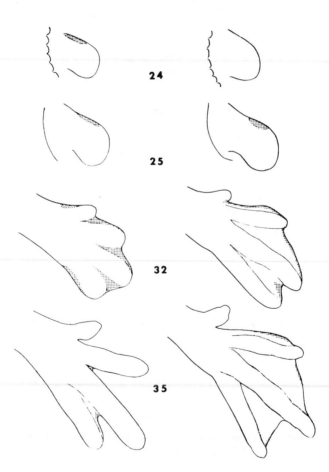

Fig. 1.

The zones homologous to those of the chick's leg show
no significant cell death. The distribution of cell
death in the duck embryo foot is correlated with the
occurrence of webbing. Between the first two digits,
which are not connected by webbing, the interdigital
zone of the foot primordia shows a v-shaped wedge of
necrosis. Between the second and third digits, however,
very little necrosis is seen. In the wing bud, however, the zones of necrosis involved in the formation
of the wing are similar in both the chick and the duck
(9,10,7).
      Processes similar to those described for the chick
also occur in the development of the foot in the rat
(11). On day 15 of gestation, there is a well developed
apical ectodermal ridge in the fetal rat's hind foot.
This ridge contains dead cells throughout its length,
but none are found in underlying mesenchyme. The marginal blood sinus is well developed. On day 16, the
apical ectodermal ridge is reduced in height and contains dead cells throughout its length. Dead cells are
also present in the interdigital mesenchyme, mainly
between the marginal sinus and the ectoderm. On day
17, the apical ectodermal ridge is further reduced in
height and contains few dead cells. At this time the
cell death in the interdigital mesenchyme is at a maximum and it occurs in a wedge shaped zone. Dead cells
are absent from the tips of digits. By day 18, the
digits are no longer webbed. Traces of the interdigital degenerating zones can sometimes be seen in small
patches of dead cells.

C. Palatal Development

      The development of the secondary palate in mammals
also involves cell death. During formation of the
secondary palate, the palatal shelves, which consist
of mesenchyme covered by a layer of epithelium, undergo
a complex reorientation from a vertical position with
the tongue between them to a horizontal position between
the tongue and the nasal septum. This fusion separates
the oral and the nasal cavities (12). The contact adhesion of the apposing epithelial cell surfaces
results in the formation of a midline epithelial seam

which undergoes degeneration and thus allows the mesenchyme of the palatine shelves to merge and form the
secondary palate (13,14,15).

D. Aortic Arches

The development of the aortic arches is an excellent example of the necessity for regression and cell
death during mrophogenesis. The arterial system in
the human as well as in the rat develops from a series
of aortic arches (12). Theoretically, six pairs of
aortic arches are formed; however, the fifth pair is
only a temporary doubling of the fourth pair. These
arches form successively and are never all present at
the same time. During development of the definitive
arterial system of the anterior region of the body, the
first and second arches regress, the third arch forms
the carotid system, the fourth arch forms, on the left
side of the body, the aortic arch, and on the right,
the subclavian, and is continuous with the seventh seg-
mental artery. The fifth arch also regresses and the
sixth arch forms the pulmonary arteries, on the right
side, and on the left, the ductus arteriosus which
closes during the neonatal period.

E. Placenta

My final example of the importance of cell degeneration during development applies not to the embryo
itself, but to the placenta. Cell death and degeneration in relation to the placenta occur in two types of
events. First, implantation of the early embryo requires the disintegration of the mucosal cells of the
uterus in preparation for the formation of the placenta
(16). Secondly, in the development of the placenta,
special cells appear, the trophoblastic giant cells,
which have been associated with programmed cell death.
These cells may be related to the separation of the
fetal placenta from the decidua at term and phagocytosis
and erosion of maternal decidua have also been attributed to these giant cells (17).

II. MECHANISMS OF CELL DEATH AND ITS CONTROL

A. The Death Program

One of the aspects of cell death that is most interesting is the control of the death program. In other words, certain cells are actually programmed to die at a specific stage in development. One might even speak, as Saunders and Fallon (7) have done, of the competence to die in response to environmental conditions which determine for other cells the response of growth or differentiation. (For similar observations in another system, see also reference 18). Saunders and Fallon (7,9) have studied a site in the chick embryo wing bud in which cataclysmic degeneration occurs with complete predictability. At stage 24 of chick development as many as 1500 to 2000 cells die during a period of 8 to 10 hours and are ingested by phagocytes. However, the cells of this zone have already differentiated their program, their internal death clock, by an earlier stage (stage 17) so that if this area is excised and grafted to a different region of the embryo, the somites, the cell die on schedule (that is, at stage 24). In contrast, cells from other parts of the embryo grafted in the place of the prospective necrotic zone develop normally. The Saunders group interpret the results as indicating the presence of "an internal death clock which ticks off necrosis at a pre-set time". If the same cells are removed at stage 17, and grafted to the region of the somites, they die on schedule, but if they are grafted to the dorsal side of the limb bud--they survive. If they are grafted later, however, at stage 22 rather than at stage 17, they die on schedule in either site. Thus, it appears that although the "death clock" is set by stage 17, it can be turned off by other factors in the environment up to stage 22. After stage 22, the death of the cells proceeds on schedule.

B. Biochemical Mechanisms

What are the mechanisms involved in cell death and its control? In many examples of cell death, lysosomes

appear to play an important role. However, the work of
Saunders and Fallon (7) and others (11,19) indicates
that in the limb, lysosomal activity is not a primary
factor. After death has already occurred, cells and
cell debris are engulfed by phagocytes.

An interesting concept arising from several investigations into cell death is that cell death is not
a passive events but an active process requiring the
programmed synthesis of specific proteins. Pratt and
Green (15) studied the inhibition of epithelial death
in rat palatal shelves grown as explants in organ culture. These structures showed cell degeneration of the
medial edges after 48 hours, whereas in the presence
of cycloheximide, a glutamine analogue which blocks
the synthesis of glycosaminoglycans and glycoproteins,
cell death was prevented. In contrast, the presence
of actinomycin-D or cytosine arabinoside, inhibitors
of RNA and DNA synthesis, did not inhibit the death of
medial edge cells. Since several lysosomal enzymes
have been reported to be glycoproteins, it is possible
that the inhibition of glycosylation of the lysosomal
enzymes or membranes may prevent synthesis of these
enzymes and result in the blocking of autolysis of the
palatal epithelial cells.

In another system, the degeneration of the uterine
epithelium during the implantation of the blastocyst,
or early embryo, there is evidence to suggest that the
degeneration of the uterine epithelium is dependent upon DNA directed messenger RNA synthesis. Finn and
Bredl (20) studied the implantation of blastocysts in
mice. If animals were given actinomycin-D, degeneration of the uterine epithelium around the blastocyst
was inhibited. Thus, it appears that the breakdown of
this uterine epithelium is controlled by information
contained in the DNA of the epithelial cells which is
released during implantation, probably by transcription
of DNA through messenger RNA.

The degenerating tail of the tadpole undergoing
metamorphosis is a classical example of cell death and
degeneration. Here too, it appears that the degeneration is not the result of intracellular lysosomal
action. However, numerous acid hydrolases increased
during the course of tail regression, but this is the

later stage, the secondary event of macrophage activity.
Although the early changes in tadpole tail regression
are not well characterized, experiments with inhibitors
of RNA and protein synthesis have shown that protein
synthesis is necessary for regression of the tail, both
in vivo and in culture (21).

Tata and his co-workers (22) have studied cell
death and degeneration in isolated tadpole tails maintained in organ culture. Control samples of isolated
tadpole tails after 8 days of culture remained the same
as their initial size. However, when triiodothyronine
was added to the medium, the tails showed visible regression and decrease in size. In contrast, with
actinomycin-D as well as triiodothyronine in the culture medium, no regression took place. The decrease in
the size of the tail cultured in vitro was accompanied
by increases in the activity of enzymes involved in regression, such as cathepsin, phosphatases and deoxyri-
bonuclease. At the same time, there was a burst of
both RNA and protein synthesis just at the time that
regression started. Inhibition of RNA synthesis with
actinomycin-D or of protein synthesis with puromycin or
cycloheximide abolished the regression of tail tissue
that was induced by the thyroid hormone (21). It seems,
therefore, that regression is not due merely to activation of existing lysosomes, but that activation of RNA
and protein synthesis is necessary, perhaps for the
formation of new hydrolase molecules. Thus, during
development cell death as well as cell growth and
maturation may require synthesis of specific proteins.
In more recent work from the same laboratory, it is
suggested that many of the initial hormone induced
changes might result from triiodothyronine induced
activation of proteolytic "cascades" (23).

Studies of the developing chick esophagus, which
also undergoes obligatory cell death, also showed an
increase in both acid phosphatase and β-glucuronidase
activity during this period of esophageal organogenesis
(24).

It is obvious that the mechanisms bringing about
cell death and its control are not well understood.
Lysosomes and lysosomal activity appear to play a role,
but whether this is primary or secondary, or even similar in all instances, is not clear. At least in some

systems, it is apparent that the phenomenon of cell
death and tissue degeneration requires active synthesis
of certain proteins, both enzymatic and structural, and
that certain cells respond with death to the same
stimuli that promote differentiation or growth in other
cells.

III. ABNORMAL CELL DEATH

If cell death and degeneration are an important
part of normal development and morphogenesis, what
happens if cell death is abnormal? Cell death may be
either excessive or insufficient, and in both cases,
the result is abnormal morphogenesis or teratogenesis.
Such effects may be observed under experimental conditions in the laboratory or in the naturally occurring
abnormal conditions produced by mutations.

A. Insufficient Cell Death

One example of a mutant in which insufficient cell
death occurs is the polydactylous mutant talpid^3 of the
chick (25). The most striking aspect of talpid^3 embryos
is abnormal limb formation with a greatly increased
number of digits. Unlike normal chick embryos, in both
the fore and hind limbs of talpid^3, massive cell death
is absent. The normal areas of dead cell groups rich
in acid phosphatase activity are missing in these mutants. This absence of cell death is thought to account
for the abnormal sculpturing of talpid^3 limbs. In the
talpid^3 embryo, the interdigital cell necrosis seen in
normal embryos is also absent so that there is very
little separation of the digits. Thus, the events that
can be analyzed in the development of the mutant form
help in understanding the morphogenetic role of cell
death in the normal genotype. The evidence from studies
with talpid^3 supports the hypothesis that interdigital
cell death plays an important role in causing separation
of the digits (26).
Another example of mutant in which there is absence
or failure of programmed cell death is the crooked neck
dwarf mutant of the chick. In this mutant, the esophagus remains occluded throughout embryonic life. The

abnormal occlusion of the esophagus is caused by genetically suppressed cell death, which as we have already
seen, is an obligatory phenomenon required to bring
about reopening of the embryonic esophagus (27). In
another part of the body, however, this mutant seems to
undergo excessive degeneration with complete deterioration of the leg musculature accompanying death of the
embryo during the last week before hatching. Paradoxi-
cally, other muscles and the internal organs of the
embryo are affected very little (28).

Abnormal degradative processes may also occur in
postnatal life. An example of this condition is osteo-
petrosis, a genetic disorder of the skeletal system
which occurs in man as well as in experimental animals.
In rats at least three mutant genes are known which
produce the condition (29,30). The disease is characterized by absence of bone remodeling so that there may
be almost a total lack of the marrow cavity while at
the same time the bone is highly susceptible to bone
fractures. Severe anemia and a marked delay in growth
are also characteristic manifestations. The influence
of other factors on the abnormal remodeling capabilities
of the bone is indicated by the observations that the
osteopetrotic condition can be cured by injection of
normal bone marrow cells or by implantation of normal
thymus glands (31,32).

B. Excessive Cell Death

As examples of mutants in which there is excessive
cell death, the mutant rumpless in chick embryos is a
classic example. It was shown by Professor Edgar
Zwilling at Storrs that the tail reduction in this
mutant results from a degeneration of the presumptive
tail tissue. The tails in rumpless are reduced or
missing because of an early destruction of the material
from which this structure is formed. This cell death
begins toward the end of the second day of incubation
and reaches its height by the end of the third day, at
the same time that in normal embryos the most active
growth of the tail begins. In the mutant embryo at
this stage, the entire indifferent mass which is the
source of material for the posterior extension of the

caudal structures is in a condition of extreme degeneration (33).

In another species, the mouse, the mutation taillessinline produces adults which in some cases do not have the slightest trace of a tail. In most cases there is a short filament consisting of skin and connective tissue only. There are also malformations of the spine, mainly spinal fusion of various kinds. The development of the tailless embryo proceeds normally up to eleven days after fertilization. At this time, a sharp constriction sets in at the proximal end of the tail. It always occurs at the same level and demarcates the tail from the trunk. From this time on, the constriction in the tail becomes more and more apparent. Thus, the tail of the mutant tailless develops quite normally up to a certain point after which there is retrogression and resorption of this structure (34).

A number of mutants which produce muscle degeneration or muscular dystrophy also seem to involve excessive degradative processes (28).

In the final example of mutant abnormality resulting from excessive cell death, the abnormality of the cell death lies in its precocious appearance. This is the case in the mutant chick called wingless, in which both wings and legs may be abnormal. The abnormal morphogenesis arises from the precocious appearance of cell death and its progressive extension beyond its normal area during the later stages when cell death also occurs in nonmutant embryos (35).

IV. EXPERIMENTAL APPROACHES

Abnormal cell death can also be produced experimentally. A classical approach for studying abnormal development is through the alteration of the environment which the developing embryo has at its disposal. This can be done by physical means such as radiation, by chemical means such as drugs or toxic elements, or by metabolic alterations which influence the embryo either directly or indirectly through their effect on the maternal organism (36,37). Another method and one which I would like to discuss in further detail

concerns the use of nutritional deficiencies or excesses during mammalian development (38).

A. Nutrient Deficiencies or Excesses

Experimental teratology actually began after the
reports by Hale in the 1930's that vitamin A deficiency
in the pregnant sow resulted in the birth of offspring
with small or missing eyes and other anomalies. This
was, in fact, the first demonstration that any environmental agent could produce gross congenital malformations in a mammal (36). Significant progress was made
in experimental studies of the teratogenic effects of
vitamin A deficiency by the work of Warkany and his
colleagues in a series of detailed and comprehensive
studies in rats. Among the defects they analyzed were
the cardiovascular abnormalities produced by vitamin A
deficiency during prenatal life. Particularly striking were anomalies of the aortic arch (39).

As I have already described, the normal development of the major arteries requires the degeneration of
significant portions of the aortic arch system of the
embryo. Figure 2 shows the hypothetic embryonic pattern of the aortic arches and the definitive pattern
of the arteries at birth. With prenatal vitamin A
deficiency, a variety of aortic arch anomalies resulted
from the incomplete regression of part of this system,
including anomalies of the right subclavian artery,
the anomaly of right aortic arch, double aortic arch,
and persistent left third aortic arch (Fig. 2).

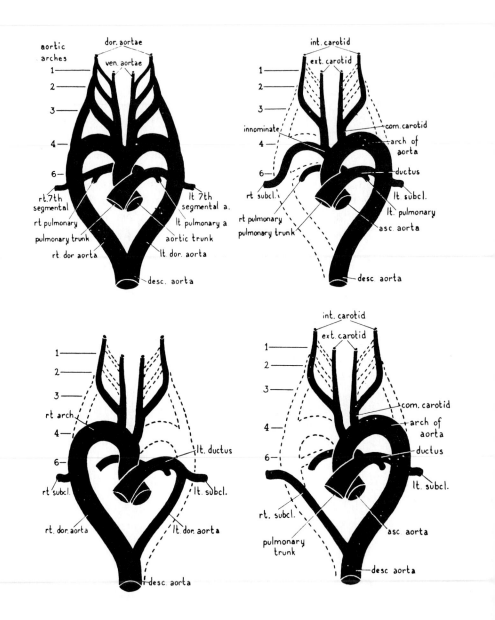

Fig. 2.

Although the mechanism of action of the teratogenicity of vitamin A deficiency remains unknown, it is clear that defective degeneration of what should be transitory structures is involved in the production of these aortic arch anomalies. This is only one example of the effects of nutritional factors in embryonic development. Many nutritional factors, especially vitamins and minerals when given in either insufficient or excessive amounts, may produce abnormal development.

B. Trace Element Deficiencies

I would now like to present some work with trace element deficiencies from my own laboratory which I believe will illustrate the approaches that may be used as tools for the study of degenerative processes in development. The first example is from a series of studies on the role of zinc in development. Zinc deficiency during prenatal development of rats produces a wide variety of congenital malformations, and these effects are produced very rapidly. When normal rats are fed a zinc deficient diet only during pregnancy, about half the implanatation sites were resorbed, and 90-100% of the full-term young were malformed. Even shorter periods of deficiency were teratogenic. The malformations affected every organ system, with gross abnormalities of the skeletal system, syndactyly, club feet, cleft lip, cleft palate, malformations of the nervous system, and of the internal organs as well (40,41, 42,43). These abnormalities were observed after only a few days of zinc deficiency. After 4 days of a zinc deficient diet, abnormal changes were already seen in the preimplantation blastocyst (44). Even at 3 days of gestation abnormal cleavage is apparent in the preimplantation egg. The rapid effect of zinc deficiency on the developing embryo stems from the speed with which the maternal level of plasma zinc fails. After only 24 hours of zinc deficiency regimen, the maternal plasma zinc level is decreased by about 40% (45). We have attempted to determine the fundamental defect underlying the abnormal morphogenesis produced by zinc deficiency. The best current hypothesis is that it is related to synthesis of nucleic acids. The uptake of tritiated thymidine in zinc deficient embryos on day 12 of gestation is much lower than normal (46). Prior injection of zinc into the mother restored the normal uptake into the embryo DNA. There also appeared to be a greater effect in the head region than in the rest of the embryo (47).

Since thymidine kinase is a zinc dependent or zinc metalloenzyme and it is required by rapidly proliferating tissues, we measured its activity in embryos from days 9 to 12 of gestation. The activity of thymidine kinase was significantly lower in zinc deficient embryos than in controls even as early as day 9 of gestation, after only 9 days of the deficient diet (48,49). Similarly the activity of DNA polymerase, a zinc metalloenzyme, was also lower in the zinc deficient embryos (49). Thus it is possible that a significant effect of zinc deficiency may be on the activity of these enzymes which are required for DNA synthesis and cell replication. However, at the same time, there is the paradoxical situation that in the esophagus there is apparently excessive proliferation with hyperkeratosis and hyperplasia of the esophageal mucosa in zinc deficient fetuses (50,51). Others have in fact shown that the rate of mitosis is higher than normal in this tissue (52).

Another effect of zinc deficiency is on the chromosomes. Chromosomal spreads prepared from fetal liver of zinc deficient fetuses and from maternal bone marrow showed chromosomal abberations especially gaps and terminal deletions, which occurred in significant incidences (53).

Thus we believe that impaired synthesis of nucleic acids caused by zinc deficiency may bring about alterations in differential rates of growth. This in turn may lead to asynchrony in histogenesis and in organogenesis. In addition there are chromosomal abberations which may also be the result of impaired synthesis of DNA.

Another possibility is that lysosomal function may play a role in the abnormal development produced by zinc deficiency. The importance of zinc for lysosomal stability has been reported by Chvapil and co-workers (54). It is possible that degradative processes involving lysosomal enzymes are abnormal, either insufficient or excessive or both, and that these might account for some of the anomalies.

Another nutrient deficiency which produces congenital abnormalities in the offspring is that of magnesium. When pregnant females are given a diet mildly deficient in magnesium there is a profound effect on fetal anemia (55). The fetuses are severely anemic although the maternal organism is normal in this respect. The anemia is characterized by edema, increased extramedullary hematopoiesis and macrocytosis, poikilocytosis, and erythroblastosis in the circulating blood. There is no evidence of alteration in the chemical structure of the hemoglobin molecule. Electron micrographs of fetal liver from controls showed erythrocytes in varying stages of maturation. Erythrocytes from magnesium deficient fetal liver showed bizarre shapes, abundance of intracellular organelles, holes or vacuoles, and disruption of the outer membrane. The morphological findings produced by magnesium deficiency in the rat fetus are consistent with the hypothesis that reduction of hemoglobin synthesis and abnormalities in the red cell membrane were the important factors in the production of the anemia.

In contrast to zinc and magnesium, a deficiency of the essential element manganese does not develop rapidly. However, manganese deficiency during both prenatal and postnatal growth affects the development of the skeleton. Disproportionate growth with short, thick bones especially of the appendicular skeleton is characteristic (56). If manganese deficiency occurs during pregnancy, the offspring exhibit congenital ataxia which results from impaired development of the inner ear (57,58). The calcified structures in the inner ear essential for the vestibular reflexes of balance, the otoliths, are imperfectly developed or absent (59,60, 61). Studies with ^{35}S radioisotope in manganese deficient animals suggest that both abnormal otolith development and abnormal bone development are related to reduced synthesis of mucopolysaccharides (62,63,64).

In attempting to establish some of the mechanisms responsible for the effects of manganese deficiency, oxidative phosphorylation was studied in isolated liver mitochondria, since metochondrial fractions contain manganese in higher proportions than do other subcellular fractions (65). In both rats and mice, the ratio of adenosine triphosphate (ATP) formed to oxygen consumed (P/O) was normal in manganese deficient animals, but oxygen uptake was reduced (66).

Electron microscope studies showed that there were ultrastructural abnormalities of liver, kidney and heart mitochondria. Some mitochondria were enlarged and often clumped, with cristae parallel to the outer membrane envelope instead of perpendicular to it (Fig. 3).

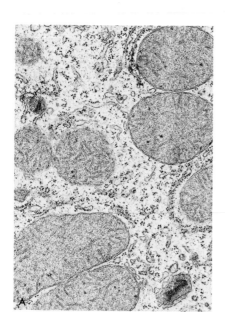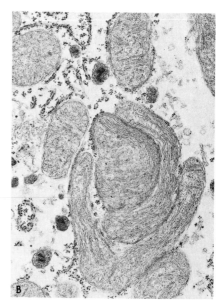

Fig. 3.

The outer mitochondrial membrane was frequently missing.
All manganese deficient tissues showed alterations in
the integrity of the cell membrane. Wherever double
membranes were found, as in the reticulum, Golgi zones,
and nuclear envelope of cells from manganese deficient
vacuoles. These changes in the Golgi and endoplasmic
reticulum are consistent with the finding of a depressed
rate of mucopolysaccharide synthesis in manganese deficient animals. It is possible that in these tissues
degradation of the membrane component of the cells is
excessive for the rate of resynthesis so that the
structural integrity of the cell membrane cannot be
maintained (67). In addition, the endoplasmic reticulum
is swollen and irregular, and microbodies occurred in
greater number than in controls. These changes are consistent with the biochemical abnormalities of reduced
oxygen uptake.

   Thus manganese deficiency produces an impaired
synthesis of mucopolysaccharides which leads to abnormal
development of otoliths, abnormal development of the
skeleton, and possibly other anomalies as well. In
addition, manganese deficiency leads to alterations in
integrity of membranes which in turn brings about
abnormal endoplasmic reticulum, abnormal Golgi, and
abnormal mitochondria. It is also possible that the
impaired mucopolysaccharide synthesis and the alterations in integrity of the membranes are related (61).

   Exactly the same kind of ataxic behavior demonstrated by manganese deficient offspring is seen in a
mutant mouse called _pallid_. Here the lack of balance
also results from missing or entirely absent otoliths
and both the abnormal morphology and the abnormal behavior can be completely prevented in the offspring
by feeding pregnant females a high level of manganese
during gestation. The demonstration that nutrient
supplementation can completely prevent the development
of a congenital genetic defect illustrates an interaction between a gene and a nutrient so that by depriving
animals of the normal genotype of manganese during their
embryonic period, a phenocopy of the mutant form is produced. In contrast, by supplementing the mutant embryo
with manganese the normal phenotype is produced (61,
59,68).

The discovery of the interaction between the mutant gene <u>pallid</u> and a trace element, manganese, led to the search for other such interactions, and we have now found two additional examples. These are mutant genes in mice in which phenotypic characteristics of the genotypes resemble those of copper deficient animals. At the same time the effects of the mutant genes can be alleviated by copper supplementation during prenatal and early postnatal life.

The <u>crinkled</u> mouse shows increased mortality in early life, thin skin with a small number of hair follicles, and abnormal texture of the hair (69). When female carriers of this mutant gene were given a stock diet or the purified control diet during pregnancy and lactation, only 22-24% of their crinkled offspring (homozygous mutant) survived to 30 days of age. In contrast, when females were fed a diet containing a high amount of copper, 50% of the crinkled offspring survived to this age. Another characteristic of this mutant is thin skin; however, when animals were given the high copper diet, skin from the mutant and nonmutant offspring was the same thickness. In addition there was an increased number of hair bulbs in the high copper fed mutants (70).

The third example that we have discovered of the interaction of a gene and a metal is the <u>quaking</u> mutant in the mouse. This mutant exhibits extensive alterations in myelination and in neural function and is characterized by tremoring. We found that the frequency of tremor was very significantly depressed in mice born to females fed a high copper diet during pregnancy and continued during lactation (71). Thus these examples demonstrate interactions in which manipulation of a single agent can produce in the wild type a phenocopy of a mutant and conversely, manipulation of the same agent can produce in the mutant the normal phenotype (60).

V. CONCLUDING REMARKS

The importance of cell death in normal morphogenesis has been emphasized. It is evident that if these degradative processes malfunction, either quantitatively or temporally, anomalous development results. The employment of mutants or of nutrient deficiencies or excesses, or the interaction of these, appears to offer useful experimental avenues to approach the study of the mechanisms involved in cell death and its control.

VI. ACKNOWLEDGEMENT

The preparation of this paper and the work from this laboratory described therein was supported in part by NIH research grants HD-01743 and HD-02355 from the National Institute of Child Health and Human Development and NIH Training Grant DE-07001 from the National Institute of Dental Research.

VII. REFERENCES

1. "The Book of Common Prayer".
2. Browning, Elizabeth Barret, A Vision of Poets.
3. Glucksmann, A., Biol. Rev. 26, 59 (1951).
4. Ham, A. W., "Histology," 7th ed., p. 378. J. B. Lippincott Company, Philadelphia, 1974.
5. Saunders, J. W., Science 154, 604 (1966).
6. Menkes, B., Sandor, S., and Ilies, A., in "Advances in Teratology" (D. H. M. Woolan, Ed.), Vol. 4, p. 169. Academic Press, New York, 1970.
7. Saunders, J. W., and Fallon, J. F., in "Major problems in developmental biology," 25th Symposium of the Society for Developmental Biology, (M. Locke, Ed.), p. 289. Academic Press, New York. 1966.
8. Pautou, M-P., J. Embryol. Exp. Morph. 34, 511 (1975).
9. Saunders, J. W., Gasseling, M. T., and Saunders, L. C., Dev. Biol. 5, 147 (1962).
10. Fallon, J. F. and Saunders, J. W., Dev. Biol. 18, 553 (1968).
11. Ballard, K. J. and Holt, S. J., J. Cell Sci. 3, 245 (1968).
12. Tuchmann-Duplessis, H., and Haegel, P., Translated by L. S. Hurley, "Illustrated Human Embryology. Vol. II. Organogenesis", Springer-Verlag,

New York, 1972.

13. Smiley, G. R., Arch. Oral Biol. 15, 287 (1970).

14. Mato, M., Smiley, G. R., and Dixon, A. D., J. Dent. Res. 5, 1451 (1972).

15. Pratt, R. M., and Greene, R. M., Dev. Biol. 54, 135 (1976).

16. El-Shershaby, A. M., and Hinchliffe, J. R., J. Embryol. Exp. Morph. 33, 1067 (1975).

17. Dorgan, W. J., and Schultz, R. L., J. Exp. Zool. 178, 497 (1968).

18. Decker, R. S., Dev. Biol. 49, 101 (1976).

19. Fallon, J. F., Brucker, R. F., and Harris, C. M., J. Cell Sci. 15, 7 (1974).

20. Finn, C. A., and Bredl, J. C. S., J. Reprod. Fert. 34, 247 (1973).

21. Tata, J. R., in "Current Topics in Developmental Biology" (A. A. Moscona, and A. Monroy, Eds.), Vol. 6, p. 100. Academic Press, New York, 1971.

22. Tata, J. R., Dev. Biol. 13, 77 (1966).

23. Smith, K. B., and Tata, J. R., Exp. Cell Res. 100, 129 (1976).

24. Wilson, J. R., and Allenspach, A. L., Dev. Biol. 41, 288 (1974).

25. Hinchliffe, J. R., and Ede, D. A., J. Embryol. Exp. Morph. 17, 385 (1967).

26. Hinchliffe, J. R., and Thorogood, P. V., J. Embryol. Exp. Morph. 31, 747 (1974).

27. Allenspach, A. L., Cytobiologie 12, 356 (1976).

28. Herrmann, H., Heywood, S. M., and Marchok, A. C. in "Current Topics in Developmental Biology" (A. A. Moscona, and A. Monroy, Eds.), Vol. 5, p. 181. Academic Press, New York, 1970.

29. Greep, R. O., J. Heredity 32, 397 (1941).

30. Moutier, R., Lamendin, H., and Berenholc, S., Experimentation animale 6, 87 (1973).

31. Milhaud, G., Labat, M-L., Graf, B., Juster, M., Balmain, N., Moutier, R., and Toyama, K., C. R. Acad. Sci. Paris 280 D, 2485 (1975).

32. Milhaud, G., Labat, M-L., Graf, B., and Thillard, M-J., C. R. Acad. Sci. Paris 283, 531 (1976).

33. Zwilling, E., Genetics 27, 641 (1942).

34. Gluecksohn-Schoenheimer, S., Genetics 23, 573 (1938).

35. Hinchliffe, J. R., and Ede, D. A., *J. Embryol. Exp. Morph.* 30, 753 (1973).

36. Kalter, H., and Warkany, J., *Physiol. Rev.*, 39, 69 (1959).

37. Wilson, J. G., and Fraser, F. C., in "Handbook of Teratology, Vol. 1. General Principles and Etiology." Plenum Press, New York, 1977.

38. Hurley, L. S. in "Handbook of Teratology" (J. G. Wilson and F. C. Fraser, Eds.), Vol. 1, p. 261. Plenum Press, New York, 1977.

39. Wilson, J. G., and Warkany, J., *Ped.* 5, 708 (1950).

40. Hurley, L. S., Gowan, J., and Swenerton, H., *Teratol.* 4, 199 (1971).

41. Hurley, L. S., and Shrader, R. E., "Neurobiology of the Trace Metals Zinc and Copper" (C. C. Pfeiffer, Ed.), Suppl. 1, p. 7. Academic Press, New York, 1972.

42. Hurley, L. S., and Mutch, P. B., *J. Nutr.* 103, 649 (1973).

43. Hurley, L. S., in "Trace Elements in Human Health and Disease, Vol. II, p. 301. Academic Press, New York, 1976.

44. Hurley, L. S., and Shrader, R. E., *Nature* 254, 427 (1975).

45. Dreosti, I. E., Tao, S., and Hurley, L. S., *Proc. Soc. Exp. Biol. Med.* 127, 169 (1968).

46. Swenerton, H., Shrader, R. E., and Hurley, L. S., *Science* 166, 1014 (1969).

47. Eckhert, C. D., and Hurley, L. S., *J. Nutr.* 107, (1977).

48. Dreosti, I. E., and Hurley, L. S., *Proc. Soc. Exp. Biol. Med.* 150, 161 (1975).

49. Duncan, J. R., and Hurley, L. S., *Fed. Proc.* 36, 1176 (1977).

50. Diamond, I., and Hurley, L. S., *J. Nutr.* 100, 325 (1970).

51. Diamond, I., Swenerton, H., and Hurley, L. S., *J. Nutr.* 101, 77 (1971).

52. Alvares, O. F., and Meyer, J., *Arch. Dermatol.* 98, 191 (1968).

53. Bell, L. T., Branstrator, M., Roux, C., and Hurley, L. S., *Teratol.* 12, 221 (1975).

54. Chvapil, M., Ryan, J., and Brada, Z., Biochem. Pharmacol. 21, 1097 (1971).
55. Cosens, G., Diamond, I., Theriault, L. L., and Hurley, L. S., Ped. Res. 11, 758 (1977).
56. Asling, C. W., and Hurley, L. S., Clin. Orthopaed. 27, 213 (1963).
57. Hurley, L. S., Everson, G. J., and Geiger, J. F., J. Nutr. 66, 309 (1958).
58. Hurley, L. S., Wooten, E., Everson, G. J., and Asling, C. W., J. Nutr. 71, 15 (1960).
59. Erway, L., Hurley, L. S., and Fraser, A., Science 152, 1766 (1966).
60. Erway, L., Hurley, L. S., and Fraser, A., J. Nutr. 100, 643 (1970).
61. Hurley, L. S., Fed. Proc. 35, 2271 (1976).
62. Hurley, L. S., Gowan, J., and Milhaud, G., Proc. Soc. Exp. Biol. Med. 130, 856 (1969).
63. Hurley, L. S., Gowan, J., and Shrader, R., in "Les Tissues Calcifies. V⁰ Symposium Europeen". Societe d'Enseignement Superieur, p. 101. Paris, France, 1968.
64. Shrader, R. E., Erway, L., and Hurley, L. S., Teratol. 8, 257 (1973).
65. Maynard, L. S., and Cotzias, G. C., J. Biol. Chem. 214, 489 (1955).
66. Hurley, L. S., Theriault, L., and Dreosti, I. E., Science 170, 1316 (1970).
67. Bell, L. T., and Hurley, L. S., Lab. Invest. 29, 723 (1973).
68. Erway, L., Fraser, A., and Hurley, L. S., Genetics 67, 97 (1971).
69. Falconer, D. S., Fraser, A. S., and King, J. W. B., J. Genet. 50, 324 (1952).
70. Hurley, L. S., and Bell, L. T., Proc. Soc. Exp. Biol. Med. 149, 830 (1975).
71. Keen, C. L., and Hurley, L. S., Science 193, 244 (1976).

AN INTEGRATED VIEW OF LYSOSOME FUNCTION

Christian de Duve

The Rockefeller University (New York)
International Institute of Cellular and Molecular
Pathology (Brussels)

Lysosomes form an intracellular digestive system. They serve in the breakdown of exogenous materials taken in by endocytosis (heterophagy), of endogenous constituents segregated from the cells' own cytoplasm (autophagy), and of secretory substances contributed by secretion granules (crinophagy). These materials are cleared from the lysosomes by digestion followed by diffusion or transport of the products across the lysosomal membrane. Non-permeant undigestible residues of this process are left behind as permanent intra-lysosomal deposits. Overloading of the lysosomes in this way may lead to storage diseases. Bulk discharge of the lysosomal contents by exocytosis (defecation) may serve as an alternative clearing mechanism, but it is restricted to a few cell types, or to pathological situations where it results in enzymatic injuries to extracellular structures.

The enzymes located in lysosomes include a wide variety of acid hydrolases acting on all major biological constituents. They are formed in the endoplasmic reticulum and transported into the lysosome system by way of the Golgi complex. With rare exceptions (neutrophils), this transfer takes place without temporary storage in special granules (primary lysosomes).

The membranes added to the lysosomes in the course of substrate and enzyme transport are removed slowly by internalization and breakdown. When rapid addition of membrane material takes place, as in endocytically active cells, much of its seems to be recycled back to the plasma membrane.

Osmotic exchanges of water between lysosomes and cytoplasm are important, sometimes leading to considerable lysosomal swelling and intense vacuolation of the cells. Protons are concentrated in lysosomes by a Donnan effect, perhaps also through the operation of a proton pump. In addition to supporting enzyme activity, protons may serve also to trap weak bases within lysosomes.

I. INTRODUCTION

The literature concerning lysosomes has swollen to considerable proportions in recent years, and a number of reviews, monographs and treatises have been devoted to the subject (1-6). To the non-specialist, such excessive information may be more bewildering than enlightening. The present paper is intended for those who wish to take a look at the forest before studying individual trees. It is restricted to a brief consideration of the main cellular mechanisms that govern lysosome function. No mention is made of how these mechanisms are adapted in each individual cell type to the particular needs of the cells and to the specific functions they perform.

II. THE CENTRAL DIGESTIVE FUNCTION OF LYSOSOMES

Lysosomes are intracellular organelles surrounded by a membrane and containing a collection of hydrolytic enzymes with an acid pH optimum. The properties of these enzymes are such that they are capable, when present together in a slightly acidic medium, of breaking down completely or almost completely all major biological constituents, including proteins, nucleic acids, lipids of various kinds, oligo- and polysaccharides, as

well as their combinations and associations. In other
words, they form a powerful digestive system.

With rare exceptions, lysosomes are actual sites
of intracellular digestion. They are digestive vacuoles, or secondary lysosomes in the jargon of lysoso-
mology. In conformity with the requirements of their
enzymes, their internal pH is acid, lying probably
somewhere between 4 and 5. The exceptions are the so-
called primary lysosomes, which serve in the transport
and storage of newly synthesized lysosomal hydrolases.
As will be mentioned below, primary lysosomes rarely
make up a significant part of a cell's lysosome complement.

As far as is known, lysosomes occur in all eukaryo-
tic cells, both in the plant and in the animal world.
Although a cell may contain as many as several hundred
of these particles, it is convenient to consider lysosomes as forming a single membrane-bounded space, which
happens to be subdivided into numerous discrete compartments as part of its particular mode of operation.
This view is not an abstraction, but is based on fact.
Lysosomes can fuse with each other to form bigger
lysosomes, just as the latter can divide into smaller
ones. Such fusion-fission processes occur more or
less continuously, causing the individual lysosomes to
exchange their contents and the composition of the whole
space to tend toward homogeneity.

For all practical purposes, therefore, we may look
at the lysosomes as a vacuolar system, within the confines of which processes of acid digestion take place.
As for any other digestive system, this activity implies
a flow of substrates, entering in complex form and leav-
ing in the form of digestion products, and a flow of
enzymes. As will be shown, these two basic flows are
crucially dependent on a third one involving membrane
material. They are associated with other movements,
especially of water and of protons. These five currents
will be considered in turn.

III. THE FLOW OF SUBSTRATES

A. Inflow

Substrates of lysosomal digestion may originate from the environment (heterophagy), from the cell's own cytoplasm (autophagy), and from secretory granules transiting through the cytoplasm (crinophagy). The three processes have in common that they depend on membrane fusion. In addition, direct permeation across the lysosomal membrane may also occur.

1. Heterophagy

Introduction of exogenous materials into the lysosome system occurs in two steps: endocytic uptake and phagosome-lysosome fusion.

a. Endocytosis. This process depends on invagination of the plasma membrane around the material to be taken in, followed by closing of the invaginated portion to form a cytoplasmic vacuole containing the engulfed material. This vacuole is known as a phagosome, pinosome or endosome.

Endocytosis can be either random or selective. In the former case, objects are taken up in simple proportion to their abundance in the extracellular medium. In the latter, they are selectively concentrated by means of binding sites present on the external face of the plasma membrane.

The size of endocytized objects can vary greatly according to cell type and circumstance, from that of small molecules (sucrose) to that of cells (microorganisms, erythrocytes). Uptake of solid particles is generally termed phagocytosis, of liquid droplets pinocytosis, and of very small objects micropinocytosis.

b. Phagosome-lysosome fusion. The usual fate of an endocytic vacuole is to fuse with a lysosome. In this process, the membranes of the two vacuoles merge to form a single continuous membrane, and their contents are mixed. The exogenous materials taken in from the environment thus enter the lysosome system and become accessible to attack by the lysosomal digestive enzymes.

This fusion process is not obligatory. For instance, in endothelial cells, phagosomes serve in vesicular transport from one cell face to the other, and are rarely intercepted by a lysosome. On the other hand, certain microorganisms, as well as certain macromolecules, can prevent the phagosome in which they are contained from fusing with a lysosome. This ability plays an important pathogenic role, by allowing the organisms or molecules taken in to escape being broken down in lysosomes.

Conversely, phagosome-lysosome fusion may occur prematurely, before closing of the endocytic invagination. This happens readily in cells attempting to engulf bulky objects or flat surfaces. As a result of this so-called "regurgitation during feeding", lysosomal enzymes are discharged into the extracellular space where they may cause damage to structural elements. This phenomenon is believed to play a role in rheumatoid arthritis and other degenerative conditions. It may be physiologically important in bone resorption.

2. Autophagy

In this process, cytoplasmic material is carried into the lysosomal spaces by an invaginated portion of the lysosomal membrane, which eventually splits off to give rise to an intralysosomal membrane-limited inclusion which is then degraded. A two-step process, in which segregation is accomplished by a non-lysosomal structure and is followed by fusion of this autophagic vacuole with a lysosome, has been implicated as well by some workers.

An important unanswered question is whether autophagy is a purely random phenomenon, or whether it can select certain cytoplasmic constituents by means of binding sites comparable to those involved in selective endocytosis, but occurring on the cytoplasmic face of the membrane. An autophagic process discriminating for damaged organelles and molecules would be particularly advantageous.

Like endocytosis, autophagy can bring into the lysosomes objects of very different sizes. Macroautophagy

is easily recognized in its early stages by the presence of entrapped organelles such as mitochondria,
endoplasmic reticulum membranes or ribosomes. Multiple
microautophagy leads to the formation of multivesicular
bodies.

3. Crinophagy

This term designates a special process of fusion
between a secretion granule and a lysosome, whereby the
contents of the granule are delivered into the lysosomal
space, while its membrane is added to the lysosomal
membrane. It has been observed in a number of endo-
crine gland cells, and serves to break down secretory
material synthesized in excess over secretion needs.

4. Permeation

In addition to the three mechanisms just described,
which have in common that they depend on processes of
membrane fusion, a variety of materials can also enter
the lysosomal space by permeation across the membrane.
This requires the molecules to be either small enough
to be able to pass through the pores of the membrane or
sufficiently lipid-soluble to traverse the lipid bilayer by phase distribution. Forming a special class
are weak bases that are permeant in uncharged form and
non-permeant in protonated form. Displacement of the
equilibrium between these two forms by the intra-
lysosomal acidity causes such compounds to become concentrated several hundred-fold within the lysosomal
space. Chloroquine and neutral red are prototypes of
this class, which comprises a large number of drugs and
dyes (7).

B. Outflow

1. Permeation after digestion

The main mechanism whereby materials entering the
lysosomal space can be cleared from it is permeation
across the membrane. As a rule, this requires preliminary digestion.

It is not known whether passage across the lysosomal membrane occurs only by simple diffusion, or may
depend also in certain cases on the participation of

carrier-mediated transport mechanisms. But there is evidence, based on the intralysosomal retention of disaccharides, and small peptides, that only very small water-soluble molecules, of molecular weight lower than about 220, can diffuse freely across the membrane. It follows that lysosomal digestion must be complete, or nearly so, for its end-products to be able to leave the lysosomal space. This is indeed what happens under normal circumstances. Substrates entering the lysosomes are broken down to such small molecules as amino acids, simple sugars, fatty acids, mononucleosides, phosphate and sulfate ions, etc., which then diffuse or are transported across the lysosomal membrane into the cytoplasm. There they are metabolized or utilized for biosynthetic purposes. In this way, vast amounts of matter may traverse the lysosomal space with nary a trace left behind. But for this to occur, there must be perfect compatibility, both quantitative and qualitative, between substrates and enzymes. If some sort of incompatibility obtains, lysosomal overloading occurs.

Numerous examples are known of this phenomenon, which is a very common pathological mechanism. One group includes all the instances in which undigestible materials, ranging from injected artificial polymers to invading microorganisms, enter the lysosomal space. Another group comprises all the conditions where the defect is on the enzyme side. Prominent in this group are most genetic storage diseases. In between are a variety of situations in which the rate at which certain substrates are taken in exceeds the rate at which they can be digested. The intralysosomal accumulation of proteins in the nephrotic kidney, and perhaps of cholesterol esters in atheromatous foam cells, are typical examples of such imbalance. In all these circumstances, undigested materials accumulate progressively within the lysosomes, with consequent overloading and enlargement of the system. The time-scale of the overloading process is very variable, from less than an hour to many years, depending on the relative rates of inflow and outflow. But eventually, it may lead to severe structural and functional alteration and even to cell death, as illustrated by the storage diseases.

Even under perfectly normal circumstances, residue accumulation with lysosomes cannot be avoided altogether. For instance, it is known that digestion-resistant materials may arise spontaneously in cells as a result of lipid peroxidation. When these materials enter lysosomes by autophagy, they contribute to the formation of lipofuscin. As a result of this phenomenon, progressive lysosome overloading is an inevitable concomitant of aging. It is particularly conspicuous in cells with a very long life-span, such as heart muscle cells or neurons.

2. Exocytic discharge

In certain cells, the lysosomal contents may be discharged extracellularly in bulk, by a process involving fusion of the lysosomal membrane with the plasma membrane. This process, called exocytosis, is essentially the same as that involved in the discharge of secretory products. When it affects lysosomes, it is called cellular defecation. Physiological in many protozoa, cellular defecation seems to be restricted in higher animals to certain cell types which either have a direct external outlet, for instance hepatocytes, or are normally implicated in the degradation of extracellular structures, for example osteoclasts. Other cells are unable to unload their lysosomes by exocytosis, or do so only under pathological conditions, in which case the released enzymes often cause damage to extracellular structural elements. This mechanism plays a significant role in the pathogeny of certain degenerative diseases. It is similar in its consequences to "regurgitation during feeding" (see above).

C. A Note On the Morphology of Lysosomes

Lysosomes are defined on the basis of two criteria: their surrounding membrane and their characteristic content in hydrolytic enzymes. These two criteria can be recognized biochemically, as they were originally, by cell fractionation studies and enzyme latency measurements. Or they can be recognized morphologically, by the cytochemical demonstration of a typical lysosomal enzyme, usually acid phosphatase, within a membrane-bounded area.

Interestingly, the usual morphological criteria of size, shape and internal structure do not apply to lysosomes, owing to their extreme variability. In fact, this variability has in itself become an identification criterion. As we now know, almost every one of the innumerable cytoplasmic inclusions described in the morphological literature as having a surrounding membrane and a polymorphic or heterogeneous content, and given such names as dense bodies, residual bodies, cytosomes, cytosegresomes, etc., are of lysosomal nature.

The morphological variability of lysosomes becomes perfectly understandable in the light of what has just been said of the substrate flow. Most of the visible contents of the lysosomal space consist of materials at various stages of digestion and of undigestible residues. When the latter are made up of a single or of only a few distinct molecular species, as in many of the storage diseases, the resulting patterns may be quite startling.

IV. THE FLOW OF ENZYMES

A. Inflow

The evidence available so far indicates that lysosomal hydrolases are manufactured in the rough-surfaced endoplasmic reticulum by membrane-bound ribosomes, and processed in the Golgi complex, from which they are channeled into the lysosome system. This process resembles that involved in secretion, with the difference that the synthesized proteins are discharged into intracellular vacuoles instead of extracellularly.

Knowledge of these mechanisms is still very fragmentary. But it appears that in many cells transport of lysosomal enzymes to their final destination follows directly upon synthesis, without an intermediary storage phase. In such cells, the primary lysosomes are represented singly by a few small vesicles transiting between the Golgi complex and the lysosomal space.

The neutrophil polymorphonuclear leucocyte is a conspicuous exception to this rule. Its primary or azurophil granules, which are of lysosomal nature--the secondary or specific granules are not--are made early in development, and subsist apparently unused until the cell has become fully mature. Their discharge occurs massively as a single event, when the cell is stimulated to engage in phagocytosis.

The situation is again reminiscent of secretion, which may or may not, depending on the cell type, involve an intermediary storage form. Gland cells in which discharge is an intermittent and regulated process have storage granules. Cells lacking this kind of regulation, such as fibroblasts or plasma cells, do not.

Whether continuous or discontinuous, the flow of enzymes into the digestive lysosomal space uses a membranous vehicle and depends on membrane fusion. It resembles the inflow of substrates in this respect.

B. Outflow

Barring exocytic discharge, which, as has been seen, is of rare occurrence, the enzymes entering the lysosomal space are cleared by digestion, in the same way as protein substrates. Little is known of the rate of turnover of the lysosomal enzymes, but the indications are that it is relatively slow. Apparently, these enzymes tend to resist acid denaturation, and thereby proteolytic attack within the lysosomes. This is to be expected on evolutionary grounds. Intra-lysosomal survival is an obvious condition of the participation of an enzymatic protein in the concerted digestive function of the lysosomes. It is a property bound to be selected for.

V. THE FLOW OF MEMBRANE

Concomitant with their acquisition of substrates and enzymes, lysosomes also gain new membrane material: from the plasma membrane in heterophagy, from secretory granules in crinophagy, and from endoplasmic reticulum or Golgi in enzyme flow and perhaps in autophagy.

Until recently, only two processes were known that could compensate for this inflow of membrane material: exocytotic discharge, with insertion of the lysosomal membrane into the plasma membrane; and interiorization, especially by microautophagy, followed by digestion. However, the former is a rare phenomenon, as has already been pointed out, and the latter is much too slow to compensate for a rapid flow of membrane material, such as occurs, for instance, in actively endocytizing cells (8).

In our Belgian laboratory, Schneider (9) has recently obtained evidence that much of the membrane material added from the plasma membrane to lysosomes during endocytosis is returned to the plasma membrane. Bulk discharge of lysosomal contents does not accompany this phenomenon, but if the lysosomes happen to contain components capable of binding strongly to the external face of the plasma membrane, these may be carried to outside by the returning piece of membrane. How this shuttle operates is not known, but the indications are that the membrane material is added back to the plasma membrane in the form of intact patches, possibly empty vesicles that fuse with the plasma membrane as in exocytosis.

Here again, some similarities with secretion are observed. There is evidence in several secretory cells that the membrane material added to the plasma membrane upon exocytic discharge of secretion granules is retrieved by an endocytic type of phenomenon, which somehow makes it available for the packaging of new secretory material.

VI. THE FLOW OF WATER

Lysosomes maintain osmotic equilibrium with the surrounding cytoplasm. This may entail considerable movements of water, as well as changes in volume, whenever digestion which increases osmotic pressure, and outflow of digestion products which decreases it, fail to keep pace with each other.

Particularly striking increases in lysosomal water and volume, manifested by intense vacuolation of the

cells, can be observed under two kinds of circumstances: a) when the cells pinocytize large amounts
of non-digestible substance that is at the same time
large enough to be retained within the lysosomes and
small enough to exert a considerable osmotic effect,
for example sucrose; b) when the cells are exposed to
certain weak bases that become highly concentrated in
lysosomes by the proton-trapping mechanisms referred to
above, for example cloroquine.

Such "swelling" of the lysosomes obviously requires
readjustment of their membrane, either through additional material originating from other compartments, or
through lysosome-lysosome fusion allowing a larger
volume to be enclosed by the same membrane surface area.
Conversely, lysosome "shrinking" must involve membrane
removal or lysosomal fission.

VII.   THE FLOW OF PROTONS

The intra-lysosomal pH is undoubtedly acid, probably lying somewhere between 4 and 5. But the
mechanism whereby this acidity is maintained is not
known with certainty. There is good evidence that it
depends at least partly on a Donnan effect produced by
fixed negative charges inside the lysosomes. Digestion, which on the whole is an acidifying process, may
also contribute to it. Whether an energy-dependent
proton pump is also implicated is still uncertain.
Arguing in support of the existence of such a pump are
the enormous amounts of weak bases lysosomes are able
to accumulate in a short time by proton trapping (7).
On the other hand, convincing proof of the operation of
a lysosomal proton pump is still lacking.

VIII.   DYNAMICS AND REGULATION

Although it is still blurred in certain areas, the
picture of lysosomal functions sketched out in the
preceding pages may now be considered fairly well
established. It rests on a wealth of convincing and
mutually confirmatory experimental evidence. But it is

essentially superficial and descriptive. Our knowledge
of how the observed processes are brought about,
powered, controlled and regulated is still abysmally
small.

The key mechanism, as in secretion and other instances of vesicular transport, are those that determine membrane movement and fusion. Understanding of
these mechanisms is one of the major challenges for
future research.

Another unsolved problem concerns the intracellular containment of lysosomal digestion. We know from
the evidence of autophagy that the lysosomal membrane
is essentially digestible by the lysosomal enzymes.
Nevertheless, it is not attacked when it serves to
separate the lysosomal space from the cytoplasm. Presumably, when in contact with normal cytoplasm, it
maintains on its inner face a molecular configuration
that is resistant to digestion. It would be particularly interesting to know the nature of the mechanism
involved, since its breakdown with consequent rupture
of the lysosomal membrane and intracellular discharge
of the lysosomal enzymes is believed to be a significant cause of cell injury.

IX. REFERENCES

1. de Duve, C., and Wattiaux, R., Functions of Lyso-
    somes, <u>Ann. Rev. Physiol</u>. 28, 435 (1966).
2. Dingle, J. T., and Fell, H. B., editors, Lysosomes
    in Biology and Pathology, Vols. 1 and 2, North-
    Holland Publishing Co., Amsterdam, 1969.
3. Dingle, J. T., editor, Lysosomes in Biology and
    Pathology, Vol. 3, North-Holland Publishing Co.,
    Amsterdam, 1973.
4. Dingle, J. T., and Dean, R. T., editors, Lysosomes
    in Biology and Pathology, Vol. 4, North-Holland
    Publishing Co., Amsterdam, 1975.
5. Hers, H. G., and Van Hoof, F., editors, Lysosomes
    and Storage Diseases, Academic Press, New York,
    1973.
6. Holtzman, E., Lysosomes: A Survey, Springer-
    Verlag, Wien New York, 1976.

7. de Duve, C., de Barsy, T., Poole, B., Trouet, A.,
    Tulkens, P., and Van Hoof, F., Biochem. Pharmacol.
    23, 2495 (1974).
8. Steinman, R. M., Brodie, S. E., and Cohn, Z. A.,
    J. Cell Biol., 68, 665 (1976).
9. Schneider, Y.-J., Etude du Sort de la Membrane
    Plasmique au Cours de l'Endocytose. Thèse doc-
    torale, Universite' Catholique de Louvain, 1976.

ELUCIDATION OF CLINICAL LYSOSOME DEFICIENCIES

Roscoe O. Brady

*National Institute of Neurological and
Communicative Disorders and Stroke
Bethesda, Maryland*

*A new era of practical medical genetics has
resulted from successful searches for enzyme deficiencies in hereditary metabolic disorders. Reliable
procedures for the diagnosis of patients, the detection of heterozygous carriers, and the prenatal
diagnosis of all of the sphingolipid storage diseases
of man are in widespread use at this time. Similar
techniques for detecting patients and carriers of
mucopolysaccharidoses and glycogenoses are under
development. Furthermore, there are strong indications that enzyme replacement may become a useful
therapeutic modality in the near future for certain
lysosomal deficiency disorders.*

The saga of precise genetic counseling for hereditary diseases has evolved over the short period of two
decades. At the end of the 19th century, clinicians
began to describe patients with deterioration of the
nervous system that was frequently accompanied by some
degree of organomegaly. Chief among these conditions
were disorders that became known as Tay-Sachs disease,
Gaucher's disease, Fabry's disease and Niemann-Pick
disease. The cause of these conditions was usually
erroneously deduced in the earliest descriptions. Relatively little new information appeared concerning
their etiology until 1934 and 1935 when Aghion (1)

DISEASE	SIGNS AND SYMPTOMS
FARBER'S DISEASE	HOARSENESS, DERMATITIS SKELETAL DEFORMATION, MENTAL RETARDATION
GAUCHER'S DISEASE	SPLEEN AND LIVER ENLARGEMENT EROSION OF LONG BONES AND PELVIS, MENTAL RETARDATION ONLY IN INFANTILE FORM
NIEMANN-PICK DISEASE	LIVER AND SPLEEN ENLARGEMENT MENTAL RETARDATION, ABOUT 30 PERCENT WITH RED SPOT IN RETINA
KRABBE'S DISEASE (GLOBOID LEUKODYSTROPHY)	MENTAL RETARDATION, ALMOST TOTAL ABSENCE OF MYELIN, GLOBOID BODIES IN WHITE MATTER OF BRAIN
METACHROMATIC LEUKODYSTROPHY	MENTAL RETARDATION, PSYCHOLOGICAL DISTURBANCES IN ADULT FORM, NERVES STAIN YELLOW-BROWN WITH CRESYL VIOLET DYE
FABRY'S DISEASE	REDDISH-PURPLE SKIN RASH, KIDNEY FAILURE, PAIN IN LOWER EXTREMITIES
TAY-SACHS DISEASE	MENTAL RETARDATION, RED SPOT IN RETINA, BLINDNESS, MUSCULAR WEAKNESS
TAY-SACHS VARIANT	SAME AS TAY-SACHS DISEASE BUT PROGRESSING MORE RAPIDLY
GENERALIZED GANGLIOSIDOSIS	MENTAL RETARDATION, LIVER ENLARGEMENT, SKELETAL DEFORMITIES, ABOUT 50 PERCENT WITH RED SPOT IN RETINA
FUCOSIDOSIS	CEREBRAL DEGENERATION, MUSCLE SPASTICITY, THICK SKIN

Fig. 1. Principal manifestations, stored lipids, and metabolic defects in the lipid storage diseases.

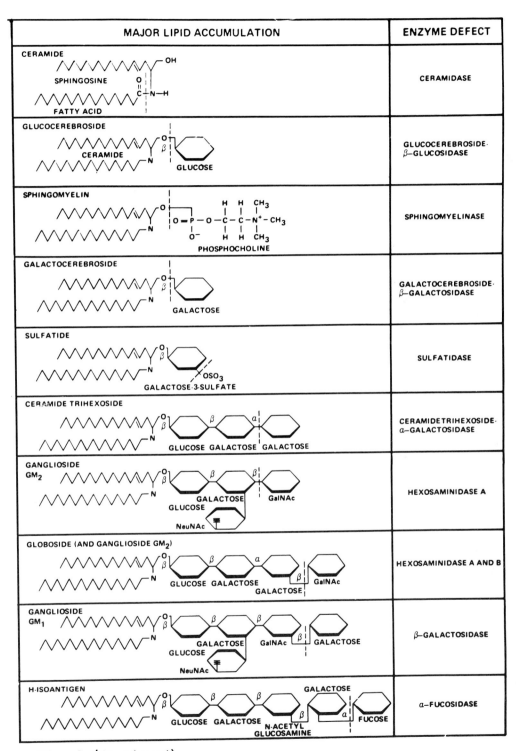

Fig. 1 (Continued)

showed that glucocerebroside (Fig. 1, line 2)
accumulated in organs and tissue of patients with
Gaucher's disease and Klenk (2) demonstrated that excessive quantities of sphinogomyelin were found in
Niemann-Pick disease (Fig. 1, line 3). The specific
compound that accumulated in Tay-Sachs disease was not
identified until 1962 (3). Excessive ceramidetrihexoside was found in tissues of patients with Fabry's
disease the following year (4).

Metabolic investigations in lipid storage diseases
began about mid-century. In 1957, E. G. Trams and I
carried out a series of experiments that provided
important clues concerning the enzymatic lesions in
Gaucher's disease and the other lipid storage disorders.
At that time, nothing was known about the metabolism of
the accumulating glucocerebroside. Because of considerable experience in fatty acid and cholesterol synthesis in tissue slice preparations, we decided to
determine whether slices of surviving spleen tissue obtained from Gaucher's disease patients could utilize
labeled precursors such as ^{14}C-glucose, ^{14}C-galactose,
or ^{14}C-sodium acetate for glucocerebroside synthesis.
We were primarily interested in learning whether such
tissues from patients could catalyze the synthesis of
galactocerebroside since an abnormality of carbohydrate
(galactose) metabolism might result in the formation of
a putatively abnormal molecule such as glucocerebroside.
We also wished to learn whether glucocerebroside, if it
could be shown to be a normal minor tissue component,
were made excessively rapidly by Gaucher patients. Both
of these irregularities were ruled out by the tissue
slice investigations and in 1960 we proposed that a defect of glucocerebroside catabolism might be the metabolic abnormality in Gaucher's disease (5). In order to
examine this possibility, we prepared glucocerebroside
labeled with radiocarbon-^{14}C in the hexose moiety by
chemical synthesis and in another preparation, with ^{14}C
in the fatty acid portion of the molecule. When we investigated the metabolism of these labeled compounds, we
quickly learned that all mammalian tissues contain an
enzyme that catlyzes the hydrolytic cleavage of the
glucose portion of the molecule (6) (Reaction 1).

1. Glucocerebroside + H_2O <u>glucocerebrosidase</u>→
    glucose + ceramide (Fig. 1, line 1)

Once the optimal conditions for measuring the activity of glucocerebrosidase were established, we assayed its activity in control human tissue specimens and in comparable samples obtained for patients with Gaucher's disease. A marked reduction of glucocerebrosidase activity was observed in all of the Gaucher preparations (7,8). This metabolic lesion in patients with Gaucher's disease has been amply confirmed.

Emboldened by this observation, we carried out similar investigations with ^{14}C-sphingomyelin labeled in the phosphocholine portion of the molecule with control human tissues and specimens obtained from patients with Niemann-Pick disease. A deficiency of the enzyme that catalyzes the hydrolytic cleavage of phosphocholine was shown to be the metabolic defect in this disorder (9) (Reaction 2).

2. Sphingomyelin + H_2O <u>sphingomyelinase</u>→ phosphocholine + ceramide

At this point, the concept of defects of lipid catabolism in sphingolipid storage diseases appeared to be on a firm foundation. On the basis of this information, the specific enzymatic lesions in Fabry's disease, Tay-Sachs disease, and generalized gangliosidosis were predicted (10). The deficiency of ceramidetrihexosidase in Fabry's disease was documented the following year (11), and in 1969, the defect in ganglioside catabolism in Tay-Sachs disease was confirmed (12). Meanwhile, in 1968, Okada and O'Brien demonstrated the accuracy of the prediction of a deficiency of ganglioside-β-galactosidase activity in patients with generalized gangliosidosis (13).

At about this time, the fusion of metabolic defects in the sphingolipidoses with the classic concept of lysosomal disorders of de Duve was demonstrated. Weinreb and co-workers (14) found that glucocerebrosidase and galactocerebrosidase (the latter enzyme was later shown to be involved in Krabbe's disease (15) (Fig. 1, line 4)) were lysosomal enzymes. Sphingomyelinase also appeared in highest specific activity in

lysosomes; however, considerably more sphingomyelinase
activity was found in other subcellular fractions than
was observed for the cerebrosidases or acid phosphatase, another typical lysosomal enzyme. The unusual
distribution of sphingomyelinase may account for some
of the variations in the clinical manifestations in
patients in the various categories of Niemann-Pick
disease (16).

Thus, the stage was set: The precedent of deficiencies of lipid-catabolizing enzymes was well established, and in a few short years, the specific metabolic defects in all of the ten known sphingolipid
storage diseases were identified (Fig. 1). Investigators concerned with management of these disorders
quickly realized that these findings might be utilized
for the development of facile diagnostic tests. The
first step in this direction was the demonstration that
washed leukocyte preparations obtained from small
samples of venous blood could be used for the diagnosis
of Gaucher's disease and Niemann-Pick disease (17).
Within a few years, the procedure had been refined so
that heterozygotes as well as affected individuals
could be identified (18). Similar assays with extracts
of cultured skin fibroblasts are widely used at the
present time (19,20). Because the activity of the
sphingolipid hydrolases is considerably greater in
these cells than in leukocytes, they are presently the
preferred source of tissue for these diagnostic tests.
Furthermore, the frozen fibroblasts may be shipped
without loss of activity from any part of the world to
an appropriate facility for testing.

The next logical development was the perfection of
similar procedures utilizing cultured fetal cells obtained by transabdominal amniocentesis for the antenatal diagnosis of the lipid storage diseases (21-24).
This major accomplishment has afforded the means for
the control of the incidence of these devastating disorders and these developments represent the epitome
of continuity of information flow between basic and
applied research.

Most of the tests chosen for the illustration of
these diagnostic reactions involved the use of authen-
tic radioactively labeled lipids as substrates.

In 1969, Okada and O'Brien showed that most patients with Tay-Sachs disease were deficient in a particular hexosaminidase isozyme (hexosaminidase A) whose activity could be measured with artificial substrates such as 4-methyl-β-$\underline{D}$-N-acetylglucosaminide (25) (Fig. 2).

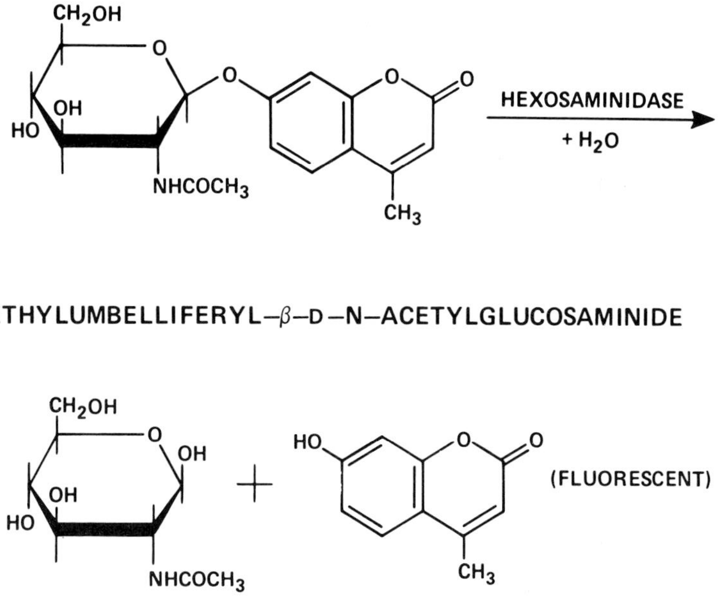

4—METHYLUMBELLIFERYL—β—D—N—ACETYLGLUCOSAMINIDE

N-ACETYLGLUCOSAMINE   4-METHYLUMBELLIFERONE

Fig. 2.   Substrate and products of the reaction for the fluorometric assay of hexosaminidase activity.

Assays with this material have been useful for the detection of Tay-Sachs homozygotes and heterozygotes (26) and it is frequently used for the prenatal diagnosis of this disorder (21,27). However, the extended use of such a substrate must be regarded with scepticism since it is now known that there are certain perfectly normal individuals with virtually no detectable hexosaminidase A activity in their tissues when measured with this fluorogenic substrate (28,29). In spite of this

finding, the catabolism of Tay-Sachs ganglioside
(ganglioside G_{M2}) is not impaired in these individuals
(30), an observation consistent with their healthy
state. Thus, if one were to use only the fluorogenic
substrate for antenatal diagnosis, such individuals
would be incorrectly identified as Tay-Sachs homozygotes.

On the other hand, it is quite clear that general
clinical chemistry laboratories are usually unable to
carry out diagnostic assays with radioactive sphingo-
lipids because of the expense of the labeled substrates
and the cost of necessary equipment. It is therefore
desirable to have simpler procedures using chromogenic
or fluorogenic substances made available to them provided the reliability of such tests is established.
Within the past two years important strides have been
made in this direction. Following the demonstration
of a deficiency of sphingomyelinase in Niemann-Pick
disease, it was reasoned that a chromogenic model compound could be developed with sufficient resemblance
to the natural lipid that it could be used as a diagnostic reagent (18) (Fig. 3).

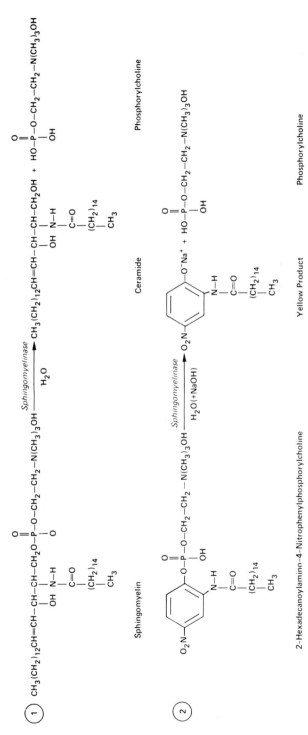

Fig. 3. The enzymatic hydrolysis of sphingomyelin (1) and the chromogenic analogue of sphingomyelin (2).

This chromogenic analogue was synthesized (31) and it
has been shown to be completely reliable for the diagnosis of Niemann-Pick homozygotes and heterozygotes
(32) as well as for the prenatal detection of Niemann-
Pick disease (33). More recently, a similar analogue
of galactocerebroside was prepared and it, too, has
been found useful for similar determinations in Krabbe's
disease (34). Thus, except for the caveat indicated for
the prenatal detection of Tay-Sachs disease, diagnostic
tests for all of the sphingolipidoses are available and
within the scope of competence of any clinical chemistry laboratory. These developments represent major
advances in the delivery of practical health care from
basic scientific endeavors.

In sum, a remarkable series of achievements has
occurred within the last dozen years regarding the
ability to control the incidence of the devastating
afflictions known as the lipid storage diseases. Progress has also recently been made in elucidating the
metabolic defects in other heritable disorders such as
the mucopolysaccharidoses and glycogenoses (35,36).
Further investigations are still required for the development of simple, useful diagnostic reagents for the
majority of these conditions. In view of the pessimis-
tic situation just over a decade ago concerning the
sphingolipidoses and the remarkable accomplishments
that have since occurred, I believe it is justifiable
to predict that similar advances will soon occur in
these disorders as well.

It seems important at this juncture to devote some
consideration to another frontier in hereditary
diseases; namely, the development of therapeutic
measures that may become useful to the treatment of
patients with these disorders. It was early recognized
that one of the most reasonable approaches along this
line would be to try to replace the missing enzyme by
administration of purified materials derived from appropriate sources (10). The magnitude of this task was
also clearly recognized, and the rather crude alternative of organ transplantation was cited as a potential
interim measure for some of the lipid storage diseases.
This procedure has been attempted in three of these
conditions. The first was a patient with the juvenile

form of Gaucher's disease who received a spleen allograph (37). The recipient manifested severe intolerance
to the grafted organ and there was no indication of any
beneficial effect. The second case was a kidney transplant in a patient with the infantile form of Gaucher's
disease (38). Here, too, there was no appreciable improvement. Both patients died within a year following
the operation. Another attempt along this line has been
carried out in a patient with the classic infantile form
of Niemann-Pick disease who received a liver allograph
(39). In this case, the investigators believed that
following the operation there was improvement in the
seizure pattern and therefore in the brain damage;
however, no analytical data have been presented to sub-
stantiate such a claim and the infant apparently has
since expired.

More extensive investigations have been carried out
along this line in patients with Fabry's disease. The
majority of these individuals suffer from renal failure
in their mid to late forties. Thus, it was reasonable
to investigate the effects of renal allographs in these
patients. Several pieces of information appear to have
evolved from such investigations. 1. In most instances, the signs and symptoms of uremia were markedly
improved following transplantation. 2. In a few patients there was apparently a temporary decrease in the
level of ceramidetrihexoside (fig. 1, line 6) in the
blood. In other trials, this effect was not observed.
3. It was agreed that the grafted kidneys do not accumulate ceramidetrihexoside. 4. There are conflicting
reports concerning the possibility of an increase of
ceramidehexosidase in the circulation after transplantation. Two years ago we had an opportunity to follow
ceramidetrihexosidase activity as well as the level of
ceramidetrihexoside in the blood of a patient with
Fabry's disease who received a kidney transplant. Although the graft appeared to function well, no increase
in enzyme activity was detected and no effect on the
level of ceramidetrihexoside was observed. Similar
results were recently reported by Van den Bergh and
coworkers (40). These investigators made the further
important observation that there was no increase in α-
galactosidase activity (an indication of

ceramidetrihexosidase) in the liver of the grafted
patient. Thus, it seems fair to conclude that renal
transplantation may be indicated as a life-saving
measure in patients with Fabry's disease who are in
renal failure. However, it is highly conjectural
whether the other deleterious consequences of the disorder such as a propensity for early myocardial infarc-
tion and cerebrovascular accidents will be alleviated
or corrected by this procedure.

The prophesy that purification of the requisite
enzymes for replacement trials might be difficult,
proved to be an underestimation of the enormity of the
task. Convenient human sources had to be discovered,
entirely new techniques for enzyme isolation had to be
developed, and rigorous testing procedures to exclude
pyrogenicity and ensure sterility of the enzymes had
to be carried out. In addition, vehicles suitable for
human administration had to be devised since the
enzymes were often isolated and preserved in a non-
compatible solvent solution. In time these challenges
were met, although essentially in reverse order of the
chronology of the discoveries of the metabolic defects.
For example, the first pure enzyme obtained and tested
was hexosaminidase A which had been isolated from
human urine. The enzyme was carefully infused intra-
venously into a patient with the Sandhoff-Jatzkewitz
variant form of Tay-Sachs disease (Fig. 1, line 8) (41).
Immediately following injection, the level of hexosaminidase A in the circulation rose to the normal
value. However, the enzyme was rapidly cleared from
the blood stream and it had a half-life of approximate
ly 7.5 minutes in the circulation. Most of the enzyme
was taken up by the liver. No enzyme appeared to cross
the blood-brain barrier, a not unexpected finding, and
there was no evidence of improvement in the clinical
course of the patient. However, several additional
important observations were made in the course of this
investigation. 1. The injected hexosaminidase caused
a 43 percent decrease in blood globoside (cf. Fig. 1,
line 8) by four hours after administration of enzyme.
This neutral glycolipid is elevated in the blood in
these patients and it accumulates in various tissues
along with ganglioside G_{M2}. 2. When the amount of

hexosaminidase A activity was compared in the liver
biopsy specimens obtained before infusion and 45
minutes after injection of enzyme, it was found that
there was 4.8 times more hexosaminidase A activity in
the liver after infusion than had actually been given
to the patient. The significance of this apparent increase in enzymatic activity was not fully appreciated
until a similar augmentation of enzyme was observed in
subsequent replacement trials in patients with Fabry's
disease (vide infra).

In order to examine the direct effect of enzyme
replacement in Fabry's disease, procedures were developed for the isolation of ceramidetrihexosidase from
human placental tissue (42,43). When this enzyme was
infused into two patients with Fabry's disease, there
was a significant decrease in the elevated level of
ceramidetrihexoside in the plasma of the recipients
which was proportional to the amount of enzyme administered (44). Since most of the accumulating
ceramidetrihexoside appears to be derived from globoside in the stroma of senescent red blood cells, the
transport of this lipid by the blood from sites of
erythrocyte catabolism such as the spleen and liver to
involved tissues such as the kidneys and blood vessels
appears to play an important role in the pathogenesis of
Fabry's disease. Thus, it may be expected that a
reduction of circulating ceramidetrihexoside as a consequence of injection of ceramidetrihexosidase will have
a salutary effect on the clinical course of the disease.

Several other important observations were made during these investigations. 1. Neither patient had any
untoward reaction to the placental enzyme. When the
Fabry patients who had received the purified placental
enzyme were tested a year after receiving the enzyme,
there was no indication that the recipients had been
sensitized to the placental preparation. This finding
is in sharp contrast with the myriad difficulties encountered by patients who received organ allographs,
and it provides strong incentive for pursuing the treatment of patients with lipid storage disease by administration of purified enzymes. 2. It appears likely that
the injected enzyme exerted its catalytic effect after

it was taken up by tissues such as the liver (45).
3. The exogenous ceramidetrihexosidase may have acti-
vitated the patient's endogenous catalytically inactive enzyme. There was 3.9 times more activity in
the liver of one of the recipients than had actually
been injected (45). This finding is consistent with
the previous observation of apparent activation of
defective hexosaminidase A in the patient with Tay-
Sachs disease. The most logical explanation of these
startling findings seems to be that a monomer of
active exogenous enzyme combines with subunits of the
patient's inactive enzyme and confers catalytic activity to the patient's mutated protein (Fig. 4).

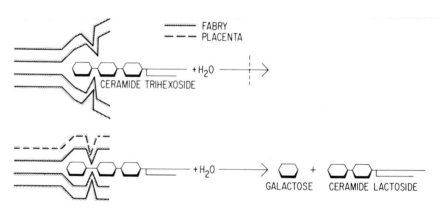

Fig. 4. Hypothetical activation of mutated catalytically inactive ceramidetrihexosidase in Fabry's
disease patients by a monomer of functional placental
enzyme.

Encouraged by the findings in Fabry's disease, we undertook a more extensive investigation of the effect of glucocerebrosidase in patients with Gaucher's disease. Procedures were developed for the isolation of this enzyme in homogenous form, again from human placental tissue (46,47). When glucocerebrosidase was infused into three patients with Gaucher's disease, there was a significant decrease in the quantity of glucocerebroside in the liver of each recipient (Table I).

TABLE I   Effect of Purified Glucocerebrosidase
on the Quantity of Glucocerebroside
in the Livers of Patients with Gaucher's Disease

Patient	Enzyme Infused	Before Infusion	24 Hours After Infusion	Glucocerebroside Change	
	units	micrograms/gram of liver		Δ	%
1	1.5×10^6	702	519	− 183	−26
2	3.3×10^6	1630	1210	− 420	−26
3	9.3×10^6	17900	16500	−1400	− 8

In addition, in two of the recipients, the elevated
level of glucocerebroside associated with circulating
erthrocytes returned to the normal level within a
period of 72 hours following infusion. This effect
of exogenous enzyme on blood glucocerebroside persisted over a comparatively long period of time
(Fig. 5) (48).

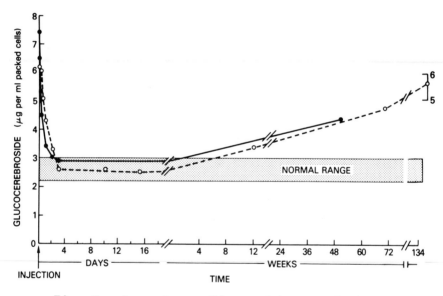

Fig. 5. Long-term effect of human placental glucocerebrosidase on the level of circulating glucocerebroside in two patients with Gaucher's disease
(Reprinted with permission from Ref. 33).

The fact that no decrease in the red cell glucocerebroside was observed in the third patient was presumably due to the extremely high level of
glucocerebroside in the liver in this individual where
there was only an 8% reduction after the administration
of enzyme. This observation is compatible with the
hypothesis that the glucocerebroside in the circulation
is a function of the amount of exchangeable glucocerebroside in tissues such as the liver (49).

Several other observations made in the course of
this investigation merit comment: 1. None of the
patients had any fever or other untoward reaction to
the injected enzyme. 2. The amount of glucocerebroside catabolized was proportional to the amount of
enzyme injected. For each unit of enzyme injected 0.38
nanomole of accumulated lipid was catabolized. The
result in the three infusions were remarkably consistent in this regard. 3. The amount of glucocerebroside cleared frcm the liver of the recipients was
equivalent to the quantity of lipid that had accumulated
over a period of 4 years in the first patient, 13 years
in the second, and 1.7 years in the third. 4. In the
post-infusion liver biopsies which were obtained 24
hours following the injection of the enzyme, no augmentation of endogenous glucocerebrosidase activity was
seen as had been observed in the investigations in Tay-
Sachs disease and Fabry's disease. Probably the best
explanation of this finding is that all of the exogenous
enzyme had been catabolized within this period of time.
The effects of its catalytic activity, which persisted
over many months, had been accomplished within this
brief time span. In subsequent investigations on the
turnover of exogenous glucocerebrosidase in rodents
and primates, a similar enzyme kinetic course was observed. A potentially important finding made in the
latter investigation was that the duration of effectiveness of the exogenous glucocerebrosidase could be
markedly extended if the animals were anaesthetized or
their body temperature was lowered (50). This discovery
may have important clinical consequences since a comparison of the Q_{10} for the enzyme with the degree of
temperature lowering permissible in patients indicates

a substantial augmentation in aggregate catalytic activity.

On the basis of these observations, I believe it is reasonable to conclude that enzyme replacement therapy in lipid storage diseases without central nervous system damage is feasible and offers a clear hope for improvement of the clinical state in such patients. In addition, one must never lose sight of the possibility that heretofore unexpected or unrelated therapeutic procedure will be discovered. A particularly intriguing aspect along this line is the possibility of activating the mutated catalytically defective enzyme in patients with such disorders. It is clear that the patients with Gaucher's disease are manufacturing some enzyme since all of these individuals have some residual glucocerebrosidase activity in their tissues. Similarly, the presence of proteins that cross-react with antibodies against normal enzymes have been shown in patients with metachromatic leukodystrophy (51), the Sandhoff-Jatzkewitz variant of Tay-Sachs disease (52) and generalized gangliosidosis (53). The potential activation of such mutated enzymes receives strong support for the work of Rotman and Celada (54) and Melchers and Messer (55) who observed that certain mutated completely inactive E. coli β-galactosidases could be made catalytically normal by the addition of antibody raised against the normal enzyme. The best interpretation of this phenomenon seems to be an imposition of an active molecular configuration on the defective enzyme. This supposition is harmonious with the hypothesis concerning the augmentation of hexosaminidase A and α-galactosidase activities observed in the patients who received these enzymes. In addition, we have observed a surprising activating effect of sodium taurocholate on the mutated glucocerebrosidase in tissues of patients with Gaucher's disease (Table II).

TABLE II Activation of Mutated
Glucocerebrosidase by Sodium Taurocholate

Experiment	Source of Cells	Taurocholate Concentration	
		Low	High
		Glucocerebrosidase (nanomoles/mg of protein/hr)	
1	Leukocytes		
	Gaucher patient	2.5	5.5
	Gaucher heterozygote	15.	18.
	Controls	22.	22.
2	Fibroblasts		
	Gaucher patient	18.	43.
	Control	190.	239.

It is believed that such an activation can be attributed to an effect of the agent on the enzyme rather than merely a reflection of detergent-like properties of these substances (56).

Another strategy that might be used to restore activity to mutated enzymes is complementation with a suitable polypeptide fragment forming a catalytically active heteropolymer (57). However, this procedure may have limited practical use since it seems likely that the mutations in various patients will be dissimilar and only a few might be expected to benefit from supplementation with any particular oligopeptide.

A further technique examined recently was the subcutaneous implantation of cultured skin fibroblasts into a patient with the Hunter syndrome, a disorder of mucopolysaccharide catabolism (58). While intervention of this type contains an element of attractiveness, it seems unlikely that it will be of general usefulness. For example, we examined the amount of glucocerebrosidase in the tissue culture medium in which normal human skin fibroblasts were grown and found an exceedingly low amount of activity, indicating little if any release of this enzyme by these cells. Furthermore, the recipients of fibroblast allographs require strong immunosuppressive measures and the potential hazardous long-term effect(s) of this type of graft are unknown.

Still other procedures have been conceived for the delivery of enzymes to patients. One method that has attracted much attention and generated an exceptional amount of enthusiasm is the encapsulation of enzymes into small lipid globules called liposomes. It is conceivable that one might want to devise such a carrier to selectively deposit the missing enzyme in a particular organ or tissue in patients with various disorders. Manipulation of the lipid composition of liposomes might be one way to achieve this goal. However, some recent disquieting data have appeared which indicate that subcellular damage may occur in tissues in animals injected with liposomes (59). Further experimentation is required to determine whether this hazard can be overcome by appropriate alterations of the lipid components of liposomes.

A potential refinement of this technique would be
the encapsulation of exogenous enzyme into red blood
cell ghosts derived from the patients own erythrocytes.
This procedure has actually been attempted in an investigation with impure placental glucocerebrosidase
in a patient with Gaucher's disease (60). Perhaps the
most significant message in the communication describ-
ing this experiment is the fact that a second larger
quantity of enzyme was administered to the same patient by direct intravenous injection. It seems reasonable to conclude that some unreported difficulty was
encountered in the erythrocyte ghost experiment.
Perhaps enzyme administered in this manner remains in
the circulation for an inordinate period of time. Bear
in mind that so far all of the evidence available indicates that exogenous enzymes exert their catalytic
effect after they have been taken up by various
tissues.

As a final consideration, I should like to return
to one of the persisting major concerns in enzyme replacement; the fact that unmodified exogenous enzymes
do not cross the blood-brain barrier. A number of
strategies have been advanced to try to overcome this
obstacle ranging from increasing the hydrophobicity
of the exogenous enzyme (61) to incorporation of
enzymes into the patient's own leukocytes (45). Although the latter procedure has in fact been success-
fully accomplished (62), many investigators are
extremely skeptical whether a sufficient quantity of
enzyme would be released within the nervous system to
improve the clinical status of the patient. However,
there is recent evidence that makes it imperative to
continue to explore the possibility of introducing
exogenous enzyme into the central nervous system. Feder
has obtained good evidence of the exchange of β-
glucuronidase between brain cells in tetraparental mice
derived from high and low β-glucuronidase strains (63).
It has been known for several years that the blood-
brain barrier can be temporarily altered by the intra-
arterial injection of hypertonic solutions (64). We
have recently observed that retrograde infusion of a
moderately hypertonic solution of mannitol into the
external carotid artery permits the penetration of a

physiologically normal amount of exogenous mannosidase
from the circulation into the brain (65). We have
further demonstrated the fact that the exogenous
mannosidase was taken up by brain cells and incorporated into subcellular elements where it might be expected
to carry out its catalytic activity on accumulated
material (66). This accomplishment is clearly a major
step forward, and if studies in primates currently
underway indicate a lack of deleterious side effects,
clinical studies may eventually be undertaken with this
and other enzymes utilizing this procedure. Although
one may anticipate unforeseen difficulties and dis-
appointments in the course of such pioneering investigations, if an indication of benefit is obtained,
it would provide strong incentive for further exploita-
tion of the procedure. The successful delivery of
exogenous enzyme to the brain of patients with central
nervous system damage is clearly one of the major
breakthroughs awaited with hopeful anticipation in the
next generation of enzyme replacement trials.

I.   REFERENCES

1. Aghion, A., LaMaladie de Gaucher dans C'enfance,
       thesis, University of Paris, (1934).
2. Klenk, E., Hoppe-Seylers Z. Physiol. Chem. 235, 24
       (1935).
3. Svennerholm, L., Biochem. Biophys. Res. Commun. 9,
       436 (1962).
4. Sweeley, C. C. and Klionsky, B., J. Biol. Chem. 238,
       PC 3148 (1963).
5. Trams, E. G., and Brady, R. O., J. Clin. Invest. 39,
       1546 (1960).
6. Brady, R. O., Kanfer, J. N. and Shapiro, D., J.
       Biol. Chem. 240, 39 (1965).
7. Brady, R. O., Kanfer, J. N., and Shapiro, D.,
       Biochem. Biophys. Res. Commun. 18, 221 (1965).
8. Brady, R. O., Kanfer, J. N., Bradley, R. M., and
       Shapiro, D., J. Clin. Invest. 45, 1112 (1966).
9. Brady, R. O., Kanfer, J. N., Mock, M. B., and
       Fredrickson, D. S., Proc. Natl. Acad. Sci. USA 55,
       366 (1966).

10. Brady, R. O., N. Engl. J. Med. 275, 312 (1966).
11. Brady, R. O., Gal, A. E., Bradley, R. M.,
    Martensson, E., Warshaw, A. L., and Laster, L.,
    N. Engl. J. Med. 276, 1163 (1967).
12. Kolodny, E. H., Brady, R. O., and Volk, B. W.,
    Biochem. Biophys. Res. Commun. 37, 526 (1969).
13. Okada, S., and O'Brien, J. S., Science 160, 1002
    (1968).
14. Weinreb, N. J., Brady, R. O., and Tappel, A. L.,
    Biochim. Biophys. Acta 159, 141 (1968).
15. Suzuki, K., and Suzuki, Y., Proc. Natl. Acad.
    Sci. USA 66, 302 (1970).
16. Brady, R. O., in "The Metabolic Basis of
    Inherited Disease" (J. B. Stanbury, J. B. Wyn-
    gaarden, and D. S. Fredrickson, Eds.) Fourth
    Edition, McGraw-Hill, New York, 1977, in press.
17. Kampine, J. P, Brady, R. O., Kanfer, J. N., Feld,
    M., and Shapiro, D., Science 155, 86 (1967).
18. Brady, R. O., Johnson, W. G., and Uhlendorf,
    B. W., Am. J. Med. 51, 423 (1971).
19. Sloan, H. R., Uhlendorf, B. W., Kanfer, J. N.,
    Brady, R. O., and Fredrickson, D. S., Biochem.
    Biophys. Res. Commun. 34, 582 (1969).
20. Ho, M. W., Seck, J., Schmidt, D., Veath, M. L.,
    Johnson, W., Brady, R. O., and O'Brien, J. S.
    Am. J. Hum. Genet. 24, 37 (1972).
21. O'Brien, J. S., Okada, S., Fillerup, D. L.,
    Veath, M. L., Adornato, B., Brenner, P., and
    Leroy, J., Science 172, 61 (1971).
22. Brady, R. O., Uhlendorf, B. W., and Jacobson,
    C. B., Science 172, 174 (1971).
23. Epstein, C. J., Brady, R. O., Schneider, E. L.,
    Bradley, R. M., and Shapiro, D., Am. J. Hum.
    Genet. 23, 533 (1971).
24. Schneider, E. L., Ellis, W. G., Brady, R. O.,
    McCulloch, J. R., and Epstein, C. J., Pediatrics
    81, 1134 (1972).
25. Okada, S. and O'Brien, J. S., Science 165, 698
    (1969).
26. O'Brien, J. S., Okada, S., Chen, A., and Fillerup,
    D. L., N. Eng. J. Med. 283, 15 (1970).
27. Schneck, L., Valenti, C., Amsterdam, O., Fried-
    land, J., Adachi, M., and Volk, B. W., Lancet I,

582 (1970).

28. Navon, R., Padeh, B., and Adam, A., Am. J. Hum. Genet. 25, 287 (1973).

29. Vidgoff, J., Buist, N., and O'Brien, J. S., Am. J. Hum. Genet. 25, 372 (1973).

30. Tallman, J. F., Brady, R. O., Navon, R., and Padeh, B., Nature 252, 254 (1974).

31. Gal, A. E., and Fash, F. J., Chem. Phys. Lipids 16, 71 (1976).

32. Gal, A. E., Brady, R. O., Hibbert, S. R., and Pentchev, P. G., N. Engl. J. Med. 293, 632 (1975).

33. Brady, R. O., Metabolism 26, 329 (1977).

34. Gal, A. E., Brady, R. O., Pentchev, P. G., Furbish, F. S., Suzuki, K., Tanaka, M., and Schneider, E. L., Clin. Chim. Acta, in press, 1977.

35. Neufeld, E. F., Lim, T. W., and Shapiro, L. J., Ann. Rev. Biochem. 44, 357 (1975).

36. Dorfman, A., and Malaton, R., Proc. Natl. Acad. Sci. USA 73, 630 (1976).

37. Groth, C. G., Hagenfeldt, S., Dreborg, S., Lofstrom, B., Ockerman, P. A., Samuelson, K., Svennerholm, L., Werner, B., and Westberg, G., Lancet I, 1260 (1971).

38. Desnick, S. J., Desnick, R. J., Brady, R. O., Pentchev, P. G., Simmons, R. L., Najarian, J. S., Swaiman, K., Sharp, H. L., and Krivit, W., in "Enzyme Therapy in Genetic Diseases: (R. J. Desnick, R. W. Bernlohr, and W. Krivit, Eds.), p. 109, The National Foundation, New York, 1973.

39. Delvin, E., Glorieux, F., Daloze, P., Gorman, J. and Block, P., Am. J. Hum. Genet. 26, 25A (1974).

40. Van den Bergh, F. A. J. T. M., Rietra, P. J. G. M., Kolk-Vegter, A. J., Bosch, E., and Tager, J. M., Acta Med. Scand. 200, 249 (1976).

41. Johnson, W. G., Desnick, R. J., Long, D. M., Sharp, H. L., Krivit, W., Brady, B., and Brady, R. O., in Enzyme Therapy in Genetic Diseases" (R. J. Desnick, R. W. Bernlohr, and W. Krivit, Eds.), p. 120. The National Foundation, New York, 1973.

42. Johnson, W. G., and Brady, R. O., in "Methods in Enzymology Vol. XXVIII." (V. Ginsburg, Ed.), p. 849, Academic Press, New York, 1972.

43. Kusiak, J. W., Quirk, J. M., and Brady, R. O., in "Methods in Enzymology" Academic Press, New York, 1978, in press.

44. Brady, R. O., Tallman, J. F., Johnson, W. G., Gal, A. E., Leahy, W. R., Quirk, J. M., and Dekaban, A. S., N. Engl. J. Med. 289, 9 (1973).

45. Brady, R. O., Pentchev, P. G., and Gal, A. E. Federation Proc. 34, 1310, (1975).

46. Pentchev, P. G., Brady, R. O., Hibbert, S. R., Gal, A. E., and Shapiro, D., J. Biol. Chem. 248, 5256 (1973).

47. Furbich, F. S., Blair, H. E., Shiloach, J., Pentchev, P. G., and Brady, R. O., in "Methods in Enzymology" Academic Press, New York, 1978, in press.

48. Pentchev, P. G., Brady, R. O., Gal, A. E., and Hibbert, S. R., J. Mol. Med. 1, 73 (1975).

49. Dawson, G., and Sweeley, C. C., J. Biol. Chem. 245, 410 (1970).

50. Pentchev, P. G., Kusiak, J. W., Barranger, J. A., Furbish, F. S., Rapaport, S. I., Massey, J. M., Brady, R. O., Metabolism, in press.

51. Stumpf, D., Neuwelt, E., Austin, J. and Kohler, P., Arch. Neurol. 25, 427 (1971).

52. Srivastava, S. K., and Beutler, E., J. Biol. Chem. 249, 2054 (1974).

53. Meisler, M., and Rattazzi, M. C., Am. J. Hum. Genet. 26, 683 (1974).

54. Rotman, B., and Celada, F., Proc. Natl. Acad. Sci. USA 60, 660 (1968).

55. Melchers, F., and Messer, W., Eur. J. Biochem. 17, 267 (1970).

56. Vahouny, G. V., Weersing, S., and Treadwell, C. R., Biochim. Biophys. Acta 98, 607 (1965).

57. Villarejo, M., Zamenhof, P. J., and Zabin, I., J. Biol. Chem. 247, 2212 (1972).

58. Dean, M. F., Muir, H., Benson, P. F., Button, L. R., Boylston, A., and Mowbray, J., Nature 261, 323 (1976).

59. Steger, L. D., and Desnick, R. J., <u>Biochim.</u>
     <u>Biophys. Acta</u> 464, 530 (1977).
60. Dale, G. L., Beutler, E., <u>Proc. Natl. Acad. Sci.</u>
     <u>USA</u> 73, 4072 (1976).
61. Brady, R. O., <u>Angewandte Chemie Intl. Ed. in</u>
     <u>English</u> 12, 1 (1973).
62. Cohen, C. M., Weissman, G., Hoffstein, S.,
     Awasthi, Y. C., and Srivastava, S. K., <u>Bio-</u>
     <u>chemistry</u> 15, 452 (1976).
63. Feder, N., <u>Nature</u> 263, 67 (1976).
64. Rapoport, S. I., and Thompson, H. K., <u>Science</u> 180,
     971 (1973).
65. Barranger, J. A., Pentchev, P. G., Rapoport, S.
     I., and Brady, R. O., <u>Proc. Natl. Acad. Sci. USA</u>,
     in press.
66. Öckerman, P.-A., <u>Lancet</u> 2, 239 (1967).

COMPLEMENT: MOLECULAR MECHANISMS, REGULATION
AND BIOLOGIC FUNCTION[1,2]

Hans J. Müller-Eberhard[3]

Research Institute of Scripps Clinic
La Jolla, California

I. INTRODUCTION

Complement (C) constitutes an integral part of the immune system. It consists of a set of proteins that occurs in plasma in inactive, but activatable form. These proteins have the potential of generating biological activity by entering into protein-protein interactions. In the course of these interactions, indigenous C enzymes are assembled and activated, and fission and fusion products of C proteins are formed. One of the fusion products is the membrane attack complex (MAC) which is responsible for the well known phenomenon of complement dependent cytolysis. Through fission products arising from enzymatic cleavage, C molecules express a plethora of different biological activities. Cells responding to the stimuli of C reaction products include polymorphonuclear leukocytes, monocytes, lymphocytes, macrophages, mast cells, smooth muscle cells and platelets. In vivo, C participates together with antibodies and various cellular elements in host defense mechanisms against infections.

[1]This is publication number 1510 from the Research Institute of Scripps Clinic.
[2]These investigations were supported by United States Public Health Service Grants AI 07007 and HL 16411.
[3]Dr. Müller-Eberhard is the Cecil H. and Ida M. Green Investigator in Medical Research, Research Institute of Scripps Clinic.

This physiological function of C has become apparent
through the recognition that genetic C deficiencies in
man are associated with increased susceptibility to infections. C also participates actively in inflammation
and immunologic tissue injury, as has been revealed
through the study of immunologic disease models in
animals and autoimmune disease in man. These seemingly
disparate biological functions of C may be reduced to
the same basic reactions on the molecular and cellular
level. Thus, C is increasingly emerging as a humoral
system capable of generating molecular effectors of
cellular functions. Background information on the
molecular biology and biochemistry of C may be obtained
from recent reviews (1-8).

II. THE PROTEINS

To date, twenty distinct proteins have been recognized as constituents of the C system: thirteen may be
considered components proper and seven regulatory proteins (Table I). The components include five proteolytic enzymes, C1r, C1s, C2, Factor B and Factor D. These
are serine proteases inasmuch as they are inhibitable
by DFP (reviewed in 9 and 10). The eight non-enzyme
components include C1q, the recognition protein of the
classical pathway; C3, the most versatile component of
the C system; C4, the modulator of C2; and C5, C6,
C7, C8 and C9, the precursors of the membrane attack
complex.

TABLE I Twenty Proteins of the Complement System

Components

Enzymes (5):   Clr, Cls, C2, B, D
Non-Enzymes (8): Clq, C3, C4, C5, C6, C7,
                  C8, C9

Regulators

Enzymes (2):   C3bINA, SCPB
Non-Enzymes (5): ClqINH, Cl INH, S-Protein,
                  β1H, P

Among the seven regulators are two enzymes. The endopeptidase C3b inactivator (C3bINA) cleaves and inactivates C3b and C4b (11,12). It is greatly enhanced in this function by β1H, its cofactor (13). The exopeptidase, serum carboxypeptidase B (SCPB), also called anaphylatoxin inactivator (AT-INA), controls the anaphylatoxin activity of the activation peptides C3a and C5a (14,15). The Clq inhibitor (ClqINH) is a little characterized protein that is capable of physically binding to Clq and thereby interfering with its functions (16,17). The Cl inhibitor (Cl INH) blocks the activity of Clr̄* and Cls̄ by firmly combining with these enzymes (18-20). The S-protein is synonymous with the membrane attack complex (MAC) inhibitor (MAC-INH) and blocks the membrane binding region of the forming MAC (21). β1H, in addition to being a cofactor of C3bINA, by itself causes active disassembly of the alternative C3/C5 convertase (13). Properdin (P) functions as stabilizer of the alternative C3/C5 convertase (22,23).

*The bar denotes the activated form of complement enzyme.

TABLE II   Proteins of the Classical Pathway
of Complement Activation

Protein	Molecular Weight	Number of Chains	Electroph. Mobility	Conc. (μg/ml)
C1q	400,000	6 x 3	$\gamma 2$	65
C1r	190,000	2	β	50
C1s	88,000	1	α	40
C2	117,000	1	β_1	25
C3	180,000	2	β_2	1600
C4	206,000	3	β_1	640

Some of the known properties of the components and
the regulatory proteins are listed in Tables II-V. The
proteins of the classical pathway of activation are
described in Table II. C1q has the largest molecular
size and is the most complex. It is a collagen-like
protein that is composed of 18 polypeptide chains containing a large number of glycine residues, hydroxypro-
line and hydroxylysine with a galactose-glucose disaccharide unit linked to many of the hydroxylysine
residues (24,25). Three similar but distinct chains
(A,B,C) of 20,000-22,000 dalton occur six times in the
molecule, forming six structural and functional subunits (25). This is apparently accomplished by the
non-covalent interaction of nine pairs of S-S linked
chains: six A-B pairs and three C-C pairs. The amino
acid sequence of the chains as well as the ultrastructure of the protein have been elucidated (26,27). C1r
consists of two apparently identical, noncovalently
linked chains (20). C1s is a single chain protein that
dimerizes in presence of Ca^{++}(28). C2 is a single chain
protein with two free sulfhydryl groups that upon oxidation with iodine form an intramolecular disulfide bond.
As a result, C2 activity in the C reaction increases up
to 20-fold (29). C2 is also the precursor of a kinin-
like peptide (30). C3, a two chain protein, is the
precursor of several biologically active fragments (1).
It supplies one of the two anaphylatoxins (C3a) (31-33),

an essential subunit (C3b) of three C enzymes, the major opsonin (C3b), the anchor of properdin (C3b) and a leukocytosis factor (C3e) (34). At least three of its physiological fragments, C3a, C3b and C3d, address specific cell surface receptors. C4 is composed of three chains, although it is synthesized as a single chain (pro-C4) and a very small amount of pro-C4 occurs also in serum (35-37). The chains of the three-chain form have been isolated and partially characterized structurally. Utilizing specific antibody to each chain it appears that the α-chain is primarily responsible for the immune adherence function of C4b and that the γ-chain serves primarily as C2a acceptor (38).

TABLE III   Proteins of the Membrane Attack
            Pathway of Complement

Protein	Molecular Weight	Number of Chains	Electroph. Mobility	Conc. (μg/ml)
C5	180,000	2	β_1	80
C6	128,000	1	β_2	75
C7	121,000	1	β_2	55
C8	154,000	3	γ_1	55
C9	79,000	1	α	60

The proteins of the membrane attack pathway are described in Table III. C5 supplies the second anaphylatoxin which is also a potent chemotactic factor (C5a) (32,39,40) and the nucleus for MAC assembly (C5b) (41). C6 and C7 are single chain proteins of similar properties. C8 is composed of three chains and may be intimately involved in the interaction of the MAC with the hydrophobic groups of membrane lipids (42). C9 is a one-chain acidic glycoprotein.

TABLE IV   Proteins of the Alternative (Properdin) Pathway
of Complement Activation

Protein	Symbol	Molecular Weight	Number of Chains	Electroph. Mobility	Conc. (μg/ml)
C3	C3	180,000	2	β	1600
Proactivator	B	93,000	1	β	200
Proactivator Convertase	D	24,000	1	α	2
C3b Inactivator	C3bINA	88,000	2	β	34
β1H	β1H	150,000	1	β	500

The proteins of the alternative pathway of C activation are described in Table IV. Factor B, or proactivator, constitutes the key enzyme of the pathway.
It is the precursor of two fragments, Ba, a chemotactic
factor (43), and Bb, the enzymatic site carrying subunit of the alternative C3/C5 convertase (9,10), and,
by itself, a macrophage and monocyte spreading factor
(44). Factor D, or proactivator convertase, is the
activating enzyme of Factor B and a trace protein in
human serum (concentration approximately 2 μg/ml) (45).
It occurs in plasma and serum only in its potentially
active form. Expression of its enzymatic activity
depends entirely on modulation of its substrate,
Factor B, by C3b. In absence of C3b, Factor D cannot
cleave Factor B. In contrast, trypsin, which can mimic
Factor D in the catalysis of C3 convertase formation,
can cleave Factor B directly, in absence of C3b (46).
It may be significant in this connection that unlike
other proteolytic enzymes, Factor D is not inhibitable
by the enzyme inhibitors of serum. Lately, the existence of an intimate relationship of Factor D to α-
thrombin has been suggested (47). However, recent
studies clearly show that Factor D and thrombin are
immunochemically unrelated proteins and that thrombin
cannot functionally substitute for Factor D (45).

C3 constitutes the major non-enzyme protein of the alternative pathway. It has been identified as the molecular equivalent of Factor A of the original description of the properdin system. Properdin itself will be discussed together with the other regulatory proteins of the complement system.

TABLE V   Regulatory Proteins of the Complement System

Protein	Symbol	Mol. Wt.	No. of Chains	Electroph. Mobility	Conc. (μg/ml)
Properdin	P	180,000	4	γ	22
β1H	β1H	150,000	1	β	530
C3b inactivator	C3bINA	80,000	2	β	34
C1q inhibitor	C1qINH	85,000	1	β	20
C$\overline{1}$ inhibitor	C$\overline{1}$ INH	105,000	1	α	240
Membrane attack complex inhibitor	MAC-INH, S	71,000	1	α	600
Antithrombin III	AT 111	64,000	1	α	230
Anaphylatoxin inactivator	AT-INA, SCPB	310,000	Multiple	α	50

Properties of the regulatory proteins are listed in Table V. Since the alternative pathway can function without properdin (P) and since P stabilizes the alternative C3/C5 convertase without otherwise changing its function, the protein is properly considered as a regulator. P is composed of four apparently identical subunits which are held together by noncovalent forces (48,49). At least two functional forms of the protein have been distinguished, native P (nP), the form found in unactivated fresh serum, and activated P. Activated P binds to particles bearing C3b on their surface, and assembles a soluble C3 convertase in the presence of

Factors B, D, C3 and Mg^{++}; nP does not. Native P is,·
however, recruited by the alternative C3/C5 convertase
through an adsorptive reaction (50). Upon enzymatic
degradation of the P-stabilized enzyme, P is in part
released as activated P. Both nP and activated P are
indistinguishable with respect to molecular size of the
subunits, electrophoretic mobility and amino acid composition (49). Both forms have the same NH_2- and COOH-
terminal amino acid sequences: NH_2-Asx-Pro-Val-Leu...
....Asp-Gly-Pro-Ser-Leu-COOH. They appear to differ in
conformation only, since they are interconvertible by
physicochemical manipulations and since nP and P may
exhibit widely differing circular dichroism spectra
(51).

     β1H is a glycoprotein which was immunochemically
defined more than ten years ago. Its powerful controlling function in the alternative pathway was recognized
only recently (52,13). This function is due to its
affinity for a strategic site on C3b (53,54). The C3b
inactivator (C3bINA) cleaves one peptide bond of the α-
chain of C3b without causing C3b to dissociate into two
fragments (12). It is an endopeptidase which is composed of two non-identical chains. The Clq inhibitor
(ClqINH) is a single chain β-globulin which can pre-
cipitate Clq in gel diffusion experiments and can
inhibit Cl action as well as the Clq-dependent cellular
cytotoxicity reaction (17). CĪ inhibitor (CĪ INH) is a
single chain acidic glycoprotein capable of blocking the
enzymic sites in the CĪ complex (19,20).

     The membrane attack complex inhibitor (MAC-INH) is
a newly recognized protein (21). It is only weakly
immunogenic and may therefore have escaped recognition
although its serum concentration is 600 μg/ml. It
inhibits the forming MAC by physically blocking its
hydrophobic membrane binding site (K_i = 39 μg/ml). A
second and apparently unrelated function of MAC-INH has
recently been pointed out. The protein significantly
decreased the rate of thrombin inactivation by antithrombin III (AT III). It does so by combining with
thrombin, not with AT III. In spite of combining with
thrombin, MAC-INH is not cleaved by thrombin and does

not inhibit the enzyme. In fact, a complex of MAC-INH, thrombin and AT III could be demonstrated in clotted plasma (55).

AT III, which is known as an efficient inhibitor of coagulation enzymes, was recently found to have the same function in the complement system as MAC-INH. It can block the membrane binding site of nascent MAC and thereby become part of the inactivated complex. Its serum concentration is 230 μg/ml and the K_i approximately 100 μg/ml (56).

The serum carboxypeptidase B (SCPB) is responsible for efficient inactivation of the anaphylatoxins and is therefore also known as anaphylatoxin inactivator (AT-INA). It is a high molecular weight enzyme which is composed of multiple subunits (14).

Several pairs of C proteins may have arisen from a common ancestral protein. First, C1r and C1s are very similar in amino acid composition and overall structure and function (2). Second, C2 (57) and factor B (58) are both linked to the HLA system and their respective genes are located in close proximity to each other on chromosome six in man (59). The enzymatic substrate and bond specificity of C2 and Factor B are identical (2,8). Third, C3 and C5 have been shown to be homologous in primary structure, at least as far as their activation peptides, C3a and C5a are concerned (60). C6 and C7 are similar in function, molecular size and amino acid composition (Podack, E. R. and Müller-Eberhard, H. J., unpublished results).

The sites of biosynthesis of C proteins are actively being explored (61). Macrophages synthesize C1q, C1r, C1s, C2, C3, C4, Factors B, D and properdin (62, 63). Fibroblasts, intestinal epithelium, peripheral mono- and lymphocytes also synthesize some of the C proteins. Highly purified human peripheral lymphocytes, for instance, were recently shown to synthesize C5 (64).

Work on the mechanism of action of the twenty C proteins has been technically possible only because each of the twenty proteins has been obtained in highly purified form, can be labeled with radioiodine without loss of biological activity and has allowed production of monospecific antiserum for recognition and quantitation. Most importantly, each protein can be quantitated by two

independent methods, a highly sensitive assay of biological activity and an immunochemical assay.

III. THE PATHWAYS

In their native form, the constituents of C enter into reversible protein-protein interactions that are in-
dicative of the molecular assembling process occurring
upon activation of the system. For example, reversible
complexes have been demonstrated for Clq, Clr, Cls and
Ca^{++}(65,66), for C2 and C4 (67), and for C5, C6, C7,
C8 and C9 (68). Similar interactions have been shown
to occur between C3 and properdin (69,70), and between
C3, Factor B and Mg^{++} (50,71). The awareness of the
possibility that physical interaction between different
protein molecules may signal that these molecules function together as a unit has greatly aided the elucidation of the molecular organization of the C pathways.
A telling example is the membrane attack complex and
its precursors.
  Upon contact with activators, the C proteins or-
ganize themselves to form two pathways of activation,
the classical and the alternative, and the common
terminal pathway of membrane attack.
  The two pathways of activation have a similar molecular organization (Fig. 1). An initial enzyme catalyzes the formation of the target bound C3 convertase
which in turn catalyzes the formation of the target
bound C5 convertase. The C5 convertase of either pathway, by cleaving C5, can set in motion the self-
assembly of the membrane attack complex, C5b-9.

The classical pathway is activated by immune complexes of the IgG and IgM type. It may also be activated
without mediation by immunoglobulins. For instance, RNA
tumor viruses (72,73), vesicular stomatitis virus (74),
certain endotoxins (75) and toxoplasma (Schreiber, R. D.
and Feldman, H., unpublished results) have been shown to
activate Cl directly. The alternative pathway is activated by plant, fungal and bacterial polysaccharides and
lipopolysaccharides in particulate form. Thus, inulin
(polyfructose), zymosan and gram negative endotoxins are
experimentally useful activators (reviewed in 6). In
addition, certain animal cells are activators of the

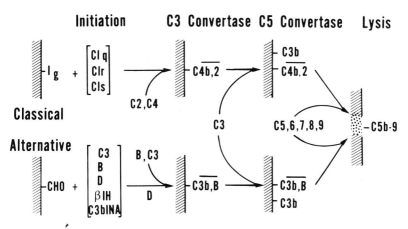

Fig. 1. Schematic representation of the classical
and alternative pathways of complement activation and of
the common, terminal pathway of membrane attack. The
shaded areas represent biological membranes on the surface of which the sequential reactions proceed. The
biologically active by-products C3a, C5a, Ba and Bb have
not been indicated. For further explanation see text.

pathway, such as lymphoblastoid cells (76) and rabbit erythrocytes (77). Although the pathway can act independently of antibodies, it may be activated by immune complexes of guinea pig γ_1 (78), aggregates of human IgA (6), or human IgG anti-viral antibody and cells infected with either measles, mumps, influenza or herpes simplex virus (79).

Several biologically important entities arise as by-products of the C reaction. Both pathways generate C3a and C5a, the two anaphylatoxins (33), and C3b, the opsonin of the C system (80,81). The classical pathway also elaborates a kinin-like peptide from C2 (30). The alternative pathway liberates Ba, the activation fragment of Factor B, which has chemotactic activity for polymorphonuclear leukocytes (43), and Bb, which is the decay product of the alternative pathway enzymes and a macrophage spreading factor (44).

It will become apparent that while most of the molecular events of the pathways are fairly well understood, the mechanisms of initiation of both pathways of activation are not clear.

IV. THE METASTABLE BINDING SITE

Although C can function in cell-free solution, it has the unusual ability to transfer itself from solution to the surface of biological particles and to function as a solid phase enzyme system. The capacity of C to mark and prepare particles for ingestion by phagocytic cells and its potential to attack and lyse cells is based on this ability of C molecules to transfer from the fluid phase to a solid phase. Transfer is accomplished through activation of metastable binding sites which are transiently revealed by the respective activating enzymes. Apparently cleavage of critical peptide bond leads to dissociation or dislocation of the activation fragment of a given component and to exposure of structures that are concealed in the native molecule. Owing to the revealed site, a molecule can bind to a suitable acceptor and establish a firm association with it. Failing collision with the acceptor within a finite time period after activation, the site decays and the

molecule remains unbound in the fluid phase. As such,
it cannot be activated again. The transfer reaction is
schematically depicted in Figure 2.

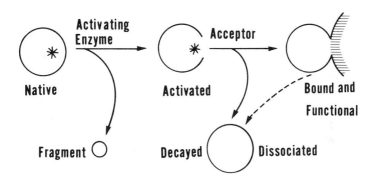

Fig. 2.   Schematic representation of activation
and decay of the metastable binding site of C2, C3, C4,
C5 and Factor B.

    The activated molecule expressing a metastable
binding site may be considered a thermodynamically un-
favorable transition form between the native and the
decayed form. The concept was first enunciated in
1965 for the binding of C4 to unsensitized erythrocytes
through the catalytic action of fluid phase C1$\overline{s}$ (82)
and it was further elaborated as "activation and transfer" hypothesis for the binding reaction of C3 (83).
Today it is clear that through this mechanism, activated C3 and C4 can bind directly to biological

membranes, activated C2 and Factor B to their respective acceptors C4b and C3b, and activated C5 to C6. An
additional metastable binding site is generated by a
non-enzymatic mechanism. Collision of the bimolecular
complex C5b,6 with C7 results in formation of the trimolecular C5b,6,7* complex which for less than ten
milliseconds has the ability to bind to biological
membranes (see below).

The rapid decay of the metastable binding sites
imposes rigorous spatial constraints on the activated
C system and prevents spreading of the effects of activation beyond the immediate microenvironment of the
activation site.

V. THE CLASSICAL PATHWAY

Basic elements of our present understanding of the
pathway have been contributed by a number of laboratories using different approaches. Mayer's group
stressed kinetic analysis, isolation of intermediate
complexes and quantitation based on the one-hit theory
(84,85). Pillemer's endeavor was directed toward isolation of C components by physicochemical methods (86).
Becker (87) as well as Lepow (88) and their associates
introduced the enzymological approach to the field.
Nelson and associates pursued functional purification
of activity entities (89). We have emphasized application of protein chemistry and analysis of molecular
mechanisms, particularly those with underlying protein-
protein interactions.

The initiation of the pathway, i.e., the internal activation of Cl, is still a puzzling problem. Cl is a Ca^{++} dependent complex of Clq, Clr and Cls, as was first shown by Lepow et al. in 1964 (65), and then further elaborated upon by Cooper (9,90) and Gigli (91,92) and their associates. In the native complex, Clr and Cls occur as inactive proenzymes. Activated Cl (C̄l) contains both enzymes in active and proteolytically cleaved form (Figs. 3 and 4). How does proteolytic cleavage come about? The complex is composed of one molecule Clq, one molecule Clr and a dimer of Cls. Clq has affinity for IgG and IgM. It has a bouquet-like ultrastructure with six flower-like globular heads which constitute the Ig binding sites (27,93). From ultracentrifugal studies it was inferred that the molecule has six or twelve independent Ig binding sites (94). The strength of binding is reflected in the single site

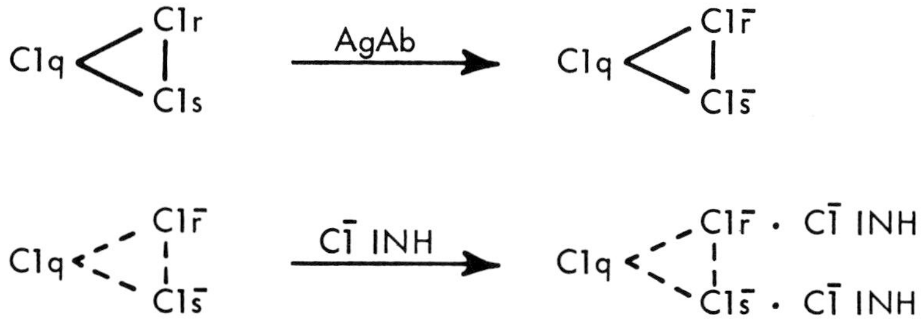

Fig. 3.   Activation and control of Cl.

molar dissociation constant. For the interactions between Clq and the subgroups of IgG this constant varies
from 2×10^{-4} to 3×10^{-5} M corresponding to a free
energy of 5 to 6 kcal/mol. Through Clq the Cl complex
can bind to immune complexes or to antibody molecules
on the surface of a target cell. If the binding is of
sufficient strength, Cl undergoes activation in the
process of which both chains of Clr and both chains of
the Cls dimer are cleaved. The resulting 60,000 and
30,000 dalton fragments remain linked by disulfide
bonds, the enzymatic sites being located in the four
light chains. Cleavage of Cls within the complex is
catalyzed by Clr̄, one molecule of enzyme being limited
to acting on only one molecule (dimer) of substrate.
Cleavage of Clr is induced by "activated" Clq. Since
Clq is not known to be an enzyme, the mechanism of
enzymatic activation of Clr remains a matter of speculation. The mechanism depicted in Figure 4 proposes
that conformationally activated Clq (Clq*) causes the
two uncleaved chains of a Clr molecule to express auto-
catalytic activity and to cleave each other. Subsequently, the cleaved molecule, Clr̄, expresses Cls cleaving activity which is independent of its interaction
with Clq*. Similarly, the enzymatic activity of Cls is
independent of the interactions between Cls̄, Clr̄ and
Clq. However, for efficient immune cytolysis the
physical integrity of the activated Cl complex is
essential.

Activation of the C1 complex renders it susceptible to the action of C1 INH. The inhibitor efficiently
binds to both C1r̄ and C1s̄, thereby totally abrogating
their enzymatic activity.
     The assembly of the classical C3 and C5 convertases
is schematically depicted in Figure 5. Attack of C4 by

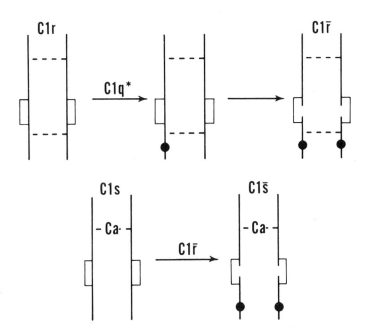

Fig. 4. Schematic representation of the mechanism
of activation of the C1 enzymes. The vertical lines
represent polypeptide chains, the dashed horizontal
lines interchain non-covalent forces, the loops intra-
chain disulfide bonds, the solid circles enzymatically
active sites.

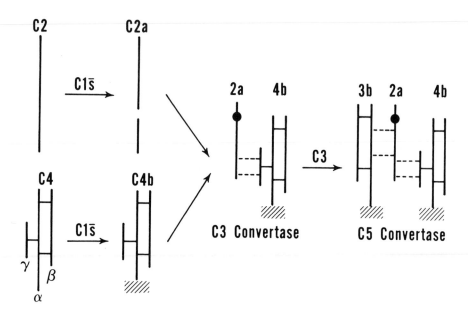

Fig. 5.   Formation of C3 and C5 convertase of the classical pathway.   The vertical lines represent polypeptide chains, the solid horizontal lines interchain covalent bonds, the dashed horizontal lines non-covalent forces, the solid circles enzymatically active sites and the shaded areas the attachment sites of, respectively, C3b and C4b on the surface of the target cell.

C1s̄ of the C1 complex activates the metastable binding site in C4 and allows C4b to attach to the target cell surface (82).   Cleavage of C2 by C1s̄ liberates the C2a fragment (85,000 dalton) and allows it to bind firmly to C4b (67).   Unlike C1r̄, C1s̄ is able to turn-over its substrates and thus to catalyze the formation of multiple C4b,2a complexes.   The γ-chain (33,000 dalton) of the three-chain C4 molecule appears to be primarily involved

in C2a binding (38). The four-chain 280,000 dalton
complex has C3 cleaving activity and is called C3 convertase. The enzymatic site of the complex is located
in the C2a subunit (95). The enzyme cleaves the bond
between residue 77 and 78 of the α-chain of C3 (96).
Dissociation of C3a (9000 dalton) allows C3b to attach
to acceptors on the surface of the target cell (83).
A C3b molecule bound in the immediate microenvironment
of a C4b,2a complex confers on this enzyme the ability
to cleave and activate C5. C5 convertase thus consists
of the six-chain 450,000 dalton C4b,2a,3b complex.
Apparently the C3b subunit contributes the function of
C5 convertase by providing the substrate binding site
(97). The enzyme cleaves the bond between residue 74
and 75 of the α-chain of C5 thus generating C5a
(12,000 dalton) and C5b which constitutes the first
intermediate product of the membrane attack pathway
(98).
      Formation and function of the enzymes are rigidly
controlled. Three regulatory mechanisms are apparent.
First, the metastable binding site of activated C4
imposes a spatial constraint on the location of the
forming enzyme. Second, decay-dissociation of C2a from
both the C3 and the C5 convertase limits their function
in time. Third, C3b and C4b can be cleaved by the
enzyme C3bINA so that they can no longer serve as
enzyme subunits.

VI. THE PROPERDIN SYSTEM

      The properdin pathway is intimately related to non-
specific defense mechanisms. In 1954 Pillemer and associates (99) formulated the first molecular concept of
natural resistance to infections according to which a
unique protein, called properdin, cooperated with
hydrazine sensitive Factor A and heat labile Factor B
in presence of Mg^{++} to achieve activation of the third
component of C and thereby killing of certain gram-
negative bacteria and neutralization of certain viruses.
Final purification of properdin from human serum was
reported in 1968 by Pensky, Hinz, Todd, Wedgwood, Boyer
and Lepow (100). Properdin was shown to be distinct

from immunoglobulins and from known C proteins. Factor
A was identified by us in 1972 as C3 (101), isolation
of which we had reported in 1960 (102). In 1971 we reported the isolation of the enzyme proactivator and
showed that upon activation it acts on C3 (103). Our
material was identified by Goodkofsky and Lepow (104) as
having the activity of Factor B. The enzyme was
originally detected through the elucidation of the
mechanism of action of cobra factor in human serum (105-
107). In 1972 we recognized a new component as an
essential part of the properdin system: proactivator
convertase, also called Factor D, which constitutes the
activating enzyme of Factor B (101). This factor was
also shown to be required in the formation of the cobra
factor dependent C3 convertase by Hunsicker, Ruddy and
Austen (108), by Cooper (107) and by Vogt, Dieminger,
Lynen and Schmidt (109). During the past three years
the importance of the C3b inactivator (12,13,22) and of
β1H (13,52,53) for the controlled function of the entire
pathway has come to light. Although considerable progress has been made in recent years toward complete
elucidation of the molecular mechanisms of the pathway,
there remain several uncertainties that require further
work.

VII. THE C3b-DEPENDENT FEEDBACK: A CHAIN REACTION

    Basic to the understanding of the alternative path-
way is the C3b initiated positive feedback mechanism
(101) (Fig. 6). A molecule of C3b and a molecule of
Factor B form, in presence of Mg^{++}, the loose bimolecular
complex C3b,B (50,71). In complex with C3b, Factor B
becomes susceptible to cleavage by Factor D, which results in formation of the C3 convertase of the alternative pathway, C3b,Bb. The a-fragment of Factor B is
dissociated in the process (109). In acting upon C3,
the enzyme supplies in a short period of time many
molecules of C3b, each of which is capable of initiating
the formation of a molecule of C3 convertase provided
the supply of Factor B is not limiting. Since Factor D

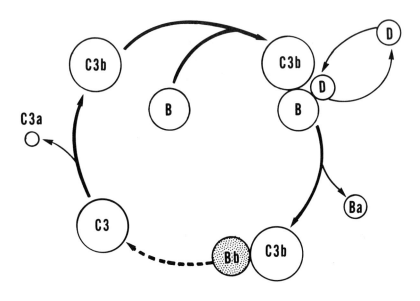

Fig. 6. The C3b-dependent positive feedback. One molecule of C3b can create many molecules of C3 convertase in presence of an unlimited supply of Factor B. Since Factor D, the activating enzyme of Factor B, is not consumed in the reaction, a chain reaction is created.

is not incorporated into the enzyme complex, it can activate many C3b,B complexes (45). In its uncontrolled form, the process resembles a chain reaction, and in presence of the regulatory proteins, a controlled chain reaction. Whereas this mechanism was first demonstrated in cell free solution (101), it was subsequently shown also to operate on the surface of cells (110).

Regulation is provided by C3bINA and by β1H. β1H binds to C3b and this binding is competitive with that of Bb (13,53). As a result, β1H disassembles C3b,Bb, dissociating Bb in inactive form. In complex with β1H, C3b is readily cleaved and inactivated by C3bINA (12, 13). With the formation of inactive C3b (C3bi), β1H is released and both control proteins are free to attack the next molecule of enzyme or of C3b. Control of formation and function of the alternative C3 convertase may be formulated as follows:

(1a) C3b + β1H $\rightleftharpoons$ C3b,β1H
(1b) $\overline{C3b,Bb}$ + β1H $\longrightarrow$ C3b,β1H + B$_i$
(2) C3b,β1H + C3bINA $\rightleftharpoons$ C3b,β1H,C3bINA
(3) C3b,β1H,C3bINA $\longrightarrow$ C3b$_i$ + β1H + C3bINA

The C3b,β1H,C3bINA complex could be demonstrated at 0°C where the apparent association constant was 10^8 M^{-1} (53).

The biomedical importance of the fact that the C3b dependent feedback is a controlled chain reaction has been pointed out by the study of genetic deficiencies of C3bINA in man. In absence of C3bINA, C3 hypercatabolism ensues which is associated with susceptibility to severe recurrent bacterial infections (111).

VIII. THE ALTERNATIVE PATHWAY

In 1976 Medicus, Schreiber, Götze and I (22) published the following concept of the molecular dynamics of the alternative pathway. According to this concept the pathway proceeds essentially in four steps (Fig. 1). First, near the surface of an activating particle the initial enzyme is assembled, which is a fluid phase C3 convertase. Assembly requires native C3, Factors B, D and Mg^{++}, but not properdin. It is the function of the enzyme to cleave C3 and to deposit C3b on the surface of the particle under attack. Second, at the site of C3b deposition, the solid phase C3 convertase (C3b,Bb) is assembled. The enzyme is a three-chain 234,000 dalton complex, its active site

being located in the Bb subunit. Third, by increasing
the multiplicity of C3b, the target bound C3 convertase
assumes C5 convertase activity. The C3/C5 convertase
constitutes a five-chain 405,000 dalton enzyme with the
Bb subunit carrying the active site (Fig. 7). Through

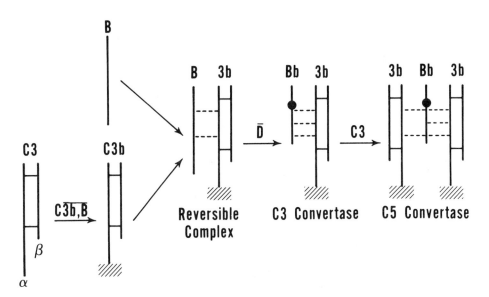

Fig. 7. Formation of the C3 and C5 convertase of
the alternative pathway. The vertical lines represent
polypeptide chains, the solid horizontal lines interchain covalent bounds, the dashed horizontal lines non-
covalent forces, the solid circles enzymatically active
sites and the shaded areas the attachment sites of C3b
molecules on the surface of the target cell.

the action of C5 convertase on C5, C-dependent cytolysis
is initiated. Fourth, properdin is recruited last in
the sequence, as a potentiating regulator of C3/C5 convertase function. Properdin is adsorbed to the solid

phase enzyme, thereby increasing the half-life of the
enzyme at 37^0C from 1.5 min. to 8-10 min. (23,50).
Although the role of properdin is not essential in the
lysis of erythrocytes via the alternative pathway (112),
it has been reported that its role is critical in the
killing of bacteria and the neutralization of certain
viruses (reviewed by Lepow (113)).

We also pointed out that there are three distinct
modes of pathway control: spontaneous decay of the
C3/C5 convertase, active disassembly of the enzyme and
inactivation of C3b by proteolytic serum enzymes (22).

Developments of the past two years have furthered
our understanding of the molecular mechanisms consti-
tuting the pathway. Whaley and Ruddy (13,54) identified
the serum protein β1H as the component responsible for
active disassembly of the C3/C5 convertase and showed
that it also accelerates C3bINA function. Pangburn
et al. (12) found an absolute requirement of β1H for inactivation of soluble C3b by C3bINA. Fearon and Austen
(114,115) showed that the C3 convertase bound to the
surface of alternative pathway activators (zymosan, rabbit erythrocytes and E. coli) is resistant to endogenous
regulatory mechanisms and that thus on the surface of
activators deregulation may prevail. In this laboratory
(53) deregulation of activator-bound C3b or C3 convertase was subsequently shown to be due to restriction of
control by β1H and it was found that such restriction of
control could be generated by chemical modification of a
non-activator (sheep erythrocytes) thereby converting it
to an activator of the alternative pathway. These advances clearly explain the known ability of activators
to sustain the "chain reaction" of C3 convertase formation in spite of the presence of the regulatory proteins.
They also shed light on the mechanism of initiation of
the pathway.

The initial molecular events are those that precede
C3b deposition and amplification of C3 convertase formation. Those events have been a subject of much experimental effort and much speculation (reviewed in ref. 6).
Invoked were immunoglobulins, properdin, or initiating
factor (IF) as ligands between the activator surface and
the precursors of the initial enzyme. Or, random deposition of C3b was envisaged as a result of spontaneous

turnover of C3 (116). In this laboratory the recognition function was assigned to what was then considered a partially purified serum protein, IF, which was difficult to separate from C3bINA (112). At that time the combined action of β1H and C3bINA as pathway regulators was not known. It is unlikely that in view of present information the IF concept of initiation is valid.

As will be presented in detail elsewhere (117), apparently five proteins are sufficient to effect initiation: C3, Factors B, D, β1H and C3bINA. Using these proteins in isolated form C3b deposition occurs on rabbit erythrocytes or zymosan in total absence of immunoglobulins, properdin and serum fractions previously referred to as IF. Since none of the five proteins individually adheres to either activator, it is concluded that the first few C3b molecules are deposited on the surface of activators by the initial C3 convertase from the fluid phase. By and large, β1H and C3bINA suppress C3 convertase formation in the fluid phase and "direct" it toward an activator because of the latter's control restricting surface (114,115). In fact, in the five component system, fluid phase C3 and Factor B consumption are almost immeasurable (117). We found initiation to be self-limiting and this phenomenon may be related to the alteration of the surface of an activator by attachment of complement molecules. When the five component system was supplemented by native properdin it became virtually indistinguishable from the alternative pathway in whole serum with respect to the ability to deposit the C3/C5 convertase on activators and to consume C3 in the fluid phase (117). It appears that after recruitment of properdin by activator-bound C3/C5 convertase, some of the properdin-stabilized enzyme dissociates in active form from the solid phase and causes fluid phase consumption of C3 (118).

Thus, immunologically, initiation of the alternative pathway appears to be non-specific. It has to do with the relative inability of β1H to contact physically activator-bound C3b. This interpretation implies that C3b has become bound to activator surface structures such that its β1H binding site is only weakly expressed or inaccessible. Whether the composition of the initial

enzyme in the fluid phase surrounding activator particles is C3b,Bb or C3,Bb, where C3 represents a rare
conformer of native C3 that functions like C3b, remains
an unresolved question.

IX. THE PATHWAY OF MEMBRANE ATTACK

     C-dependent cytolysis is entirely a function of
the activated membrane attack complex (MAC or C5b-9)
(41,119,120). The following experiment defined the
pathway as a distinct entity (121): ^{51}Cr-labeled sheep
erythrocytes were incubated in a solution of C5, C6,
C7, C8 and C9 together with cells bearing the classical
C5 convertase (EC4b,2a,3b). Not only the enzyme bearing cells, but also the previously untreated erythrocytes were lysed under these conditions, as indicated
by ^{51}Cr-release. All five proteins appeared necessary
for occurrence of significant release within a 30 min.
period. The experiment indicated that none of the
components acting prior to C5 is required on the surface of the target cell for actual membrane attack and
cell lysis. The reactive lysis system of Thompson and
Lachmann (122,123) established the same fact upon its
elucidation in molecular terms.
     The five precursor proteins of MAC form a loose
protein-protein complex (68) which upon cleavage of C5,
by either the classical or the alternative C5 convertase, self-assembles into an exceedingly firm complex;
upon dissociation of C5a, nascent C5b forms a stable
bimolecular complex with C6 (123-126). Then C5b,6 and
C7 form a trimolecular complex which for less than 9
msec has the ability to bind to the surface of biological membranes (127). Subsequently, C8 is adsorbed to
C5b-7 and C9 to C5b-8 (Fig. 8). Leakage of cell membranes ensues at the C8 stage of the assembly, indicating a major role of C8 in the interaction of MAC with
the interior of a membrane (128,129). Present evidence
suggests that C5b-7 overcomes the charge barrier and

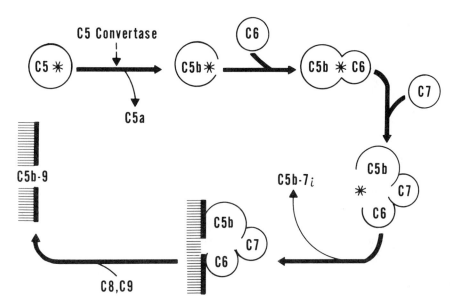

Fig. 8. Assembly of the membrane attack complex. The asterisks denote concealed and transiently revealed binding sites, and the shaded areas biological membranes.

the fully assembled MAC the hydrophobic barrier of a biological membrane (see above).

Since the assembly of MAC on a target cell proceeds in discrete steps, it was possible to probe the topology of its subunits by inhibition of MAC assembly with specific antibody molecules. For instance, treatment of C5b-7 bearing cells with antibody to C5, C6 or C7 inhibits subsequent binding of C8.

Treatment of cells bearing C5b-8 with anti-C8 inhibits
attachment of C9. It thus became apparent that C5b,
C6, C7, C8 and C9, on the surface of a target, form a
supramolecular organization in which the subunits are
in close physical proximity to each other. In 1972 we
proposed a model of the MAC according to which the trimolecular C5b-7 complex, attached to the target cell
surface, constituted the binding site for one molecule
of C8, and C8 in the tetramolecular C5b-8 complex provided the binding sites for multiple molecules of C9
(119).

The existence of the postulated complex was veri-
fied in 1973 (41). It was found that C5b-9 that does
not become attached to cells or particles remains in
cytolytically inactive form in the fluid phase (Fig. 9).
The soluble, stable complex was originally isolated from
inulin activated serum (130). The corrected sedimentation coefficient, $s^0_{20,w}$ was reported to be 28.5S, the
diffusion coefficient 1.98 x 10^{-7} cm^2/sec, the molecular
weight 1.04 x 10^6 dalton, the frictional ratio 1.48, and
the electrophoretic mobility at pH 8.6, -4.7 x 10^{-5}
cm^2 V^{-1} sec^{-1}. As predicted, the complex reacted with
antisera to C5, C6, C7, C8 and C9. It could be dissociated without reduction by sodium dodecyl sulfate
(SDS) and upon SDS polyacrylamide gel electrophoresis
the complex was found to consist of equimolar quantities
of C5b, C6, C7 and C8 and three molecules of C9 (Fig.
10). C8 was found to consist of two non-covalently

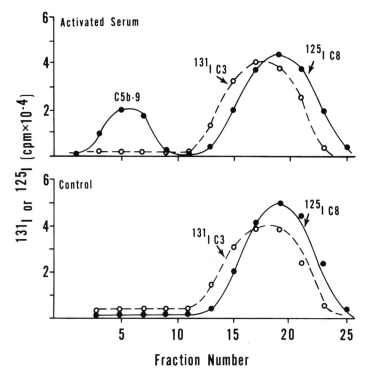

Fig. 9. Demonstration of the C5b-9 complex by sucrose density gradient ultracentrifugation of whole, inulin activated human serum supplemented with radio-labeled C3 and C8. The control was not activated. Direction of sedimentation is to the left.

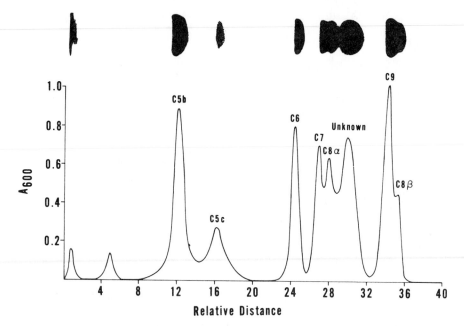

Fig. 10. Subunit composition of the isolated C5b-9 complex as analyzed by sodium dodecyl sulfate polyacrylamide gel electrophoresis. The top of the gel is at the left. The lower part shows the corresponding densitometric scan. C8a:α-γ chain of C8; unknown: S-protein.

linked subunits. In addition to the aforementioned subunits, the complex contained up to three molecules of an 88,000 dalton glycoprotein. This protein has now

been isolated from serum and identified as the major
non-lipoprotein inhibitor of the MAC (126). It is
termed the S-protein.

The MAC is an unusually firm complex. The dissociation constants for the reaction of C7 with C5b,6, of
C8 with C5b-7 and of C9 with C5b-8 are of the order of
10^{-12} M. The free energy, ΔG, of C3b-9 formation is
-50 Kcal/mol. The activation energy of C5b-7 formation
(from C5b,6 and C7) is +33.7 Kcal/mol, and the association rate constant for the same reaction is 0.14×10^5
$M^{-1} s^{-1}$ at $4^{\circ}C$ and 30 times that value at $20^{\circ}C$. The
marked temperature dependence of the association rate
constant indicates that at $20^{\circ}C$, C5b-7 formation is
diffusion controlled and at $4^{\circ}C$ it is energy controlled
(127).

Instead of binding to cell membranes, the forming
MAC may bind to plasma inhibitors. To date, three plasma
proteins are known to function as MAC inhibitors: α-
lipoprotein (126,131), S-protein (126,132) and antithrombin III (ATIII) (56). These proteins apparently
interact with the critical site of the forming MAC that
has affinity for membranes, since they effectively
compete with cell membranes for the complex.

The mode of action of the MAC on membranes has not
been completely defined. Three alternative models have
been considered: enzymatic degradation of membrane constituents, formation of a transmembrane channel, or
formation of protein-phospholipid micelles. As to the
first alternative, impairment of synthetic lipid bilayers
by complement was not accompanied by the appearance of
detectable quantities of enzymatic degradation products
of the lipids used (133). As to the second alternative,
evidence for partial insertion of C5b-9 into lipid bilayers has been obtained by proteolytic "stripping"
studies in which the components were utilized in radiolabeled form (134,135). In our laboratory it was shown
using electron paramegnetic resonance spectroscopy that
C5b-7 affects spin labels in the polar zone of flat
lipid bilayers and that C5b-8 and C5b-9 affect spin
labels located in the apolar hydrocarbon zone of such
bilayers (136). Functional studies showed that liposomes
become leaky for trapped marker after reaction with
nascent C5b-8 and that this release was enhanced by

C9 (137). Similarly, planar lipid bilayer membranes
exhibited increased ion permeability following application of C5b-8 and this increase was greatly augmented
by C9 (138). Thus, the accumulated evidence suggests
that the forming C5b-9 complex acquires a lipid binding
capacity that none of the precursor proteins possesses.
As to the third alternative, we have demonstrated previously that, as the assembly of the complex progresses,
the ability to bind certain detergents increases. C5b,6
and C5b-7 can bind 20 mol. of desoxycholate (DOC) per
mol of complex. C5b-8 and C5b-9 can bind, respectively,
60 and 80 mol DOC per mol complex (139). None of the
precursors bound significant amounts of DOC. Work in
progress indicates that these complexes also bind [14]C-
lecithin in considerable quantities (Podack, E. R., and
Müller-Eberhard, H. J., unpublished observations).

C5b-9 phospholipid complexes and micelles have been
demonstrated by ultracentrifugation and electron microscopy. The S-protein was dissociated from S-C5b-9 by
DOC in an exchange reaction and DOC was then replaced
in the DOC-C5b-9 complex by phospholipid. Using negative staining electron microscopy, the phospholipid-
C5b-9 complexes had a circular appearance measuring
200 Å in diameter and having a 100 Å diameter inner dark
circular area. This morphology bore resemblance to the
typical C-dependent ultrastructural membrane lesions
(140).

On the basis of the available body of information,
it may be proposed that the forming MAC causes ultrastructural and functional membrane impairment by binding
and thereby rearranging membrane phospholipid molecules
in its immediate microenvironment.

X.   THE MULTIFUNCTIONALITY OF C3

C3 is perhaps the key component of the entire C-
system. It has a multiplicity of functions which de-
serve to be summarized separately. C3 undergoes the
following physiological, enzymatic reactions:

$$C3 \xrightarrow{\text{C3 convertase}} C3a + C3b$$

$$C3a \xrightarrow{\text{SCPB}} C3a_i$$

$$C3b \xrightarrow[\beta IH]{\text{C3bINA}} C3b_i$$

$$C3b_i \xrightarrow{\text{tryptic enzyme}} C3c + C3d$$

$$C3c \xrightarrow{\text{tryptic enzyme}} C3(c\text{-}e) + C3e$$

As emphasized above, the action of C3 convertase of either pathway generates transiently a binding site in C3b through which C3b can firmly attach to a large variety of biological particles. C3a (9000 dalton) is one of the anaphylatoxins and is converted by serum carboxypeptidase B (SCPB) to its inactive des-Arg form (14). C3b (171,000 dalton) when bound to particles or soluble immune complexes functions as "opsonin", i.e., it marks these entities for ingestion by phagocytic cells and serves as a ligand to C3b-specific cell surface receptors (80,81,141,142). It is the acceptor of Bb and thus a subunit of the alternative C3 convertase, C3b,Bb. It constitutes an essential part of the substrate binding site of the classical and the alternative C5 convertase (reviewed in ref. 6). It serves as the receptor of activated properdin (70) and it can bind β1H. Binding of β1H results in dissociation of Bb from C3b,Bb and in conformational adaptation of C3b to the C3bINA (13,52-54). This enzyme cleaves the α-chain of C3b into a 67,000 and a 43,000 dalton fragment. Since both fragments are disulfide bonded to the β-chain, the C3bINA cleaved molecule (C3b$_i$) has the same molecular size as C3b (12). C3b$_i$ is biologically inactive. The most recently detected function of target bound C3b appears to be its ability to distinguish between an activator and a non-activator of the alternative pathway, which is reflected in a differential accessibility of bound C3b to the regulator β1H (53,114,115,117).
C3b$_i$ is highly susceptible to tryptic enzymes. In serum it is probably plasmin that cleaves its α-chain in the NH_2-terminal region and thereby severs the C3d

portion from the C3c portion of C3b$_i$ (12,143-146).
Particle-bound C3d is able to serve as ligand to C3d-
specific cell surface receptors (147). C3c is precursor of a 10,000 dalton acidic fragment, C3e, which is
probably derived from its α-chain portion. C3e is a
leukocytosis producing factor (34,148). Figure 11

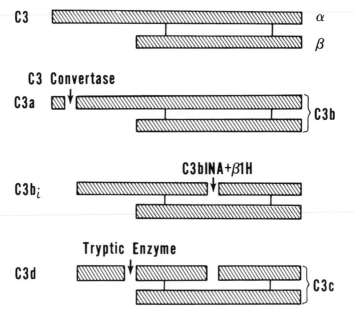

Fig. 11. Schematic description of the physiological activation and control of C3: In succession three
enzymes cause liberation of C3a anaphylatoxin, generation of C3b and its transient binding site, inactivation of C3b and fragmentation of C3b$_i$ into two immunochemically distinct pieces.

summarizes schematically the chain structure of C3 and physiological reaction products.

XI. COMPLEMENT AND INFLAMMATION

In recent years C has been increasingly implicated as a powerful humoral effector of inflammation and immunologic tissue injury. Physiological fragments of C proteins, in acting on cell surface receptors, elicit a variety of cellular responses that cause or enhance the inflammatory process. Fragments of C with biologic activity identified to date are: C3a, C5a, C3b, C3e, Ba, Bb and C2-kinin.

Evidence for the effector function of C in inflammation and tissue injury derives from a large body of in vitro and in vivo studies. Experimental diseases such as nephrotoxic nephritis (149), immune vasculitis (149), necrotizing arteritis of immune complex disease (150), myasthenia gravis (151) and ischemic myocardial necrosis (152-154) can be alleviated or prevented by temporary inactivation of circulating C. Inactivation is readily achieved by infusion of cobra venom factor, the C3b equivalent of the cobra (155) which forms a stable C3 and C5 cleaving enzyme with Factor B of the recipient (106,149). In absence of experimental C inactivation, C proteins are found deposited at the site of the tissue lesions in all of these diseases. In man, deposits of C3 and other C proteins are regularly seen in the characteristic tissue lesions of glomerulonephritis, lupus erythematosus, rheumatoid arthritis and a number of dermatological disorders (reviewed in ref. 156). Patients with diseases involving C activation usually exhibit an increased metabolism of C proteins as revealed by administration of radiolabeled C components (157-159). Finally, the intracutaneous administration of minute amounts of C3a (160,161) or C5a (15) resulted in immediate formation of edema and erythema due to degranulation of cutaneous mast cells.

The most thoroughly studied of the biologically active C fragments are human C3a and human C5a. The primary amino acid structure of C3a was elucidated by

Hugli (96) and that of C5a by Fernandez and Hugli (98).
Structural analysis became possible when methods were
devised allowing isolation of both peptides directly
from yeast activated whole human serum (15,162,163).

C3a contains 77 amino acid residues and has a
molecular weight of 9000 dalton. C5a contains 74 amino
acid residues and a sizable carbohydrate moiety which
is attached to the asparginine residue in position 64.
The molecular weight of C5a is approximately 11,200
dalton and that of its peptide portion 8200 dalton.
The NH_2-terminal of C3a is serine, that of C5a is
threonine. Both peptides have arginine in COOH-terminal
position. This arginine is absolutely essential for
the expression of the anaphylatoxin activity of both
peptides, but not for the chemotactic activity of C5a,
since C5a-desArg retains approximately 1/20 of the
activity of C5a. Upon proper alignment it became
apparent that 29 of the 74 residues of C5a are homologous to the corresponding residues of C3a (98). This
observation suggested an ancestral relationship between
C3 and C5 (60) and indicated that the similar functional properties of the two peptides are based on a
similar structure. Both peptides contain six Cys
residues in nearly identical positions, including two
repeating Cys sequences. Since these residues are
interconnected by disulfide bonds, the structures of
both peptides are restrained by a disulfide knot. C5a
contains a seventh Cys residue in position 27. Whereas
C3a has three methionine residues which are located in
the NH_2-terminal half of the sequence, C5a possesses
only one methionine residue and this occupies position
70 near the COOH-terminus. Since the COOH-terminal
sequence of C5a is important for chemotactic activity,
as evidenced by the reduced activity of C5a-desArg,
Fernandez and Hugli (98) suggest that the single methionine may be reponsible for the peptide's chemotactic
activity and explain why C3a lacks this activity. The
compactness of the structure of the two peptides is
further enhanced by a considerable content of α-helical
conformation, C3a having 40-45% (164) and C5a 50%
(165,98). Some of the properties of the two anaphylatoxins are summarized in Table VI.

TABLE VI   Properties of Human Anaphylatoxins

Properties	C3a	C5a
Electrophoretic Mobility	+2.1	-1.7
Molecular Weight:		
Total	9,000	11,200
Peptide	9,000	8,200
Carbohydrate	0	3,000
Amino Acid Residues	77	74
NH_2-Terminus	serine	threonine
COOH-Terminus	arginine	arginine
Activity:		
Ileum Contraction	1×10^{-8} M	5×10^{-10} M
Leukotaxis	inactive	3×10^{-9} M
Histamine Release	1×10^{-6} M	1×10^{-6} M
Lethal Shock (Guinea Pig)	1×10^{-8} mol	1×10^{-9} mol
Edema and Erythema	2×10^{-12} mol	1×10^{-15} mol

The history of anaphylatoxin research has been described in detail in a recent review (8). Suffice it to point out in this context the few developments that led to the present biochemical understanding of anaphylatoxins. In 1909 Friedberger (166) showed that serum after incubation with immune precipitates in vitro elicited lethal shock upon injection into animals. Vogt and associates showed that anaphylatoxin was a low molecular weight product of anaphylatoxinogen (167), a plasma protein that could be depleted in vivo by the injection of cobra venom factor (168). Vogt and associates (169) also funished the first chemical analysis of "classical" anaphylatoxin (porcine) in terms of amino acid composition and carbohydrate content. In 1967 Lepow and associates (31) showed that a low molecular weight fragment with anaphylatoxin activity is derived from human C3 as a result of C3 convertase action on the protein. In the same year, Jensen (39) found

anaphylatoxin activity to originate from C5 of guinea
pig serum. In 1968 it was shown in this laboratory
(32) that both human C3 and human C5 are precursors of
two functionally distinct anaphylatoxins, C3a and C5a.
Following chemical characterization of C3a (170), serum
carboxypeptidase B was shown to be the controlling
enzyme of C3a and C5a (14) and thus it became possible,
by selective inhibition of this enzyme, to produce
liberal amounts of active anaphylatoxins in whole human
serum (15). These developments set the stage for
structural chemistry which was followed by the synthesis of oligopeptides with anaphylatoxin activity (171).

Although not an inflammatory peptide, the C2 derived kinin is an edema producing substance that is
thought to be responsible for the clinical manifestations of patients with hereditary angioedema (30).
Present evidence indicates that the peptide is derived
from one of the C1s̄ produced C2 fragments by the action
of plasmin. The peptide is dialyzable, poor in aromatic
amino acid residues, and its activity is destroyed by
trypsin or carboxypeptidase B.

Like the C2 kinin, C3e is a secondary peptide of a
physiological fragment of C3 (34). It is liberated
from C3c by trypsin, or, in serum, by an unidentified
tryptic enzyme. C3e mobilizes leukocytes from bone
marrow and injected into rabbits (10 µg/kg) causes leukocytosis. The peptide has a molecular weight of
10,000 dalton, contains 101 amino acid residues, lacks
methionine and tyrosine and at pH 8.6 migrates upon
electrophoresis as a prealbumin (-7.5×10^{-5} cm^2 V^{-1}
sec^{-1}). C3e also causes increased vascular permeability
in rabbit skin. A complement derived leukocytosis factor was originally described by Rother (172). It is
noteworthy that an individual with homozygous C3 deficiency did not respond with leukocytosis during severe
bacterial infections including septicemia (173).

Figure 12 summarizes schematically the manner in
which biologically active C fragments may act in concert
to effect inflammation. Upon activation of either

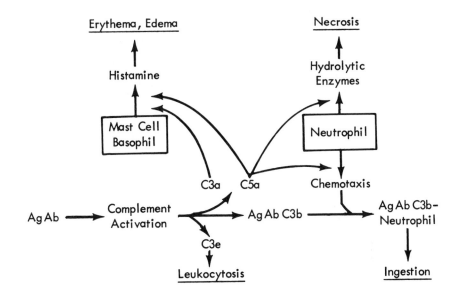

Fig. 12. Schematic representation of ligand (C3b) and messenger functions (C3a, C3e, C5a) of biologically active C fragments effecting inflammation.

pathway, C reaction products with ligand or messenger function evolve. C3b bound to immune complexes (AgAb) or to activators of the alternative pathway functions as a ligand between these substances and the specific C3b receptors on the surface of polymorphonuclear leukocytes. The activation peptides C3a and C5a cause histamine release from mastocytes or basophils and thus enhance vascular permeability, providing an increased supply of plasma proteins including C. In the skin

these events lead to formation of edema and erythema.
C5a and presumably Ba (43) address cell surface receptors of various leukocytes and subject them to directed
migration of chematactic attraction. Leukocytes arriv-
ing at the site of C activation are trapped by cell
surface receptor-C3b interaction. C5a, and to some
extent also C3a, elicit hydrolytic enzyme release from
leukocytes and thus initiate tissue necrosis. Particles
coated by C3b adhere to and are ingested by leukocytes.
The secondary peptide of C3c, C3e, causes systemic
leukocytosis.

XII. CONCLUSION

An up to date review of diseases associated with
genetic C deficiences in man is at hand (174). The
elaborate molecular mechanisms described above are, in
all probability, very intimately involved in the defense of the host against infectious agents. There is
no question about C's involvement in immunologic tissue
injury and inflammation as such. Experimental diseases
in animals and their prevention by C inactivation in
vivo have amply proven this fact. Future work should
describe the defined molecular mechanisms underlying
biologic phenomena in biochemical terms. Particularly,
the manner in which the various C reaction products
affect cellular functions is deservant of much exploration. As examples to spur investigative efforts may
serve C5a, the chemotaxin and anaphylatoxin, and Bb,
the macrophage-monocyte spreading factor. Of premier
importance for medicine, however, is to discover or
develop agents that can selectively interfere with the
inflammation producing products of the C system.
Diseases like lupus erythematosus, rheumatoid arthritis,
glomerulonephritis and ischemic myocardial necrosis may
then be alleviated by effective anticomplement treatment.

XIII. REFERENCES

1. Müller-Eberhard, H. J., Ann. Rev. Biochem. 44, 697
    (1975).

2. Cooper, N. R., and Ziccardi, R. J., in "Proteolysis and Physiological Regulations" (D. W. Ribbons, and K. Breed, Eds.), Miami Winter Symposia 11, p. 167. Academic Press, New York, 1976.
3. Müller-Eberhard, H. J., in "Textbook of Immunopathology" (P. A. Miescher, and H. J. Muller-Eberhard, Eds.), Second Edition, p. 45. Grune and Stratton, New York, 1976.
4. Osler, A. G., "Foundations of Immunology Series". Prentice-Hall, Inc., Englewood Cliffs, New Jersey, 1976.
5. Reid, K. B. M., and Porter, R. R., in "Contemporary Topics in Molecular Immunology" (F. P. Inman, and W. J. Mandy, Eds.), Vol. 4, p. 1. Plenum Press, New York, 1975.
6. Götze, O., and Müller-Eberhard, H. J., Adv. Immunol. 24, 1 (1976).
7. "Comprehensive Immunology. Biological Amplification Systems in Immunology" (N. K. Day, and R. A. Good, Eds.), Vol. 2. Plenum Publishing Co., New York, 1977.
8. Hugli, T. E., and Müller-Eberhard, H. J., Adv. Immunol., Vol. 26, in press (1978).
9. Cooper, N. R., and Ziccardi, R. J., J. Immunol. 119, 1664 (1977).
10. Medicus, R. G., Götze, O., and Müller-Eberhard, H. J., Scand. J. Immunol. 5, 1049 (1976).
11. Cooper, N. R., J. Exp. Med. 141, 890 (1975).
12. Pangburn, M. K., Schreiber, R. D., and Müller-Eberhard, H. J., J. Exp. Med. 146, 257 (1977).
13. Whaley, K., and Ruddy, S., J. Exp. Med. 144, 1147 (1976).
14. Bokisch, V. A., and Müller-Eberhard, H. J., J. Clin. Invest. 49, 2427 (1970).
15. Vallota, E. H., and Müller-Eberhard, H. J., J. Exp. Med. 137, 1109 (1973).
16. Conradie, J. D., Volanakis, J. E., and Stroud, R. M., Immunochemistry 12, 967 (1975).
17. Ghebrehiwet, B., and Müller-Eberhard, H. J., J. Immunol. 120, 27 (1978).
18. Landerman, N. S., Webster, M. E., Becker, E. L., and Ratcliffe, H. E., J. Allergy 33, 330 (1962).

19. Harpel, P. C., and Cooper, N. R., J. Clin. Invest. 55, 593 (1975).

20. Ziccardi, R. J., and Cooper, N. R., J. Immunol. 116, 504 (1976).

21. Podack, E. R., Kolb, W. P., and Müller-Eberhard, H. J., Fed. Proc. 36, 1209 (1977).

22. Medicus, R. G., Schreiber, R. D., Götze, O., and Müller-Eberhard, H. J., Proc. Natl. Acad. Sci. USA 73, 612 (1976).

23. Fearon, D. T., and Austen, K. F., J. Exp. Med. 142, 856 (1975).

24. Calcott, M. A., and Müller-Eberhard, H. J., Biochemistry 11, 3443 (1972).

25. Porter, R. R., Fed. Proc. 36, 2191 (1977).

26. Reid, K. B. M., and Porter, R. R., Biochem. J. 155, 19 (1976).

27. Knobel, H. R., Villegar, W., and Isliker, H., Europ. J. Immunol. 5, 78 (1975).

28. Valet, G., and Cooper, N. R., J. Immunol. 112, 339 (1974).

29. Polley, M. J., and Müller-Eberhard, H. J., Exp. Med. 126, 1013 (1967).

30. Donaldson, V. H., Rosen, F. S., and Bing, D. H., Trans. Assoc. Amer. Physicians 90, 174 (1977).

31. Dias Da Silva, W., Eisele, J. W., and Lepow, I. H., J. Exp. Med. 126, 1027 (1967).

32. Cochrane, C. G., and Müller-Eberhard, H. J., J. Exp. Med. 127, 371 (1968).

33. Müller-Eberhard, H. J., in "Molecular and Biological Aspects of the Acute Allergic Reactions" (S.G.O. Johansson, K. Strandberg, and B. Uvnäs, Eds.), Nobel Symposium 33, p. 339. Plenum Publishing Co., New York, 1976.

34. Ghebrehiwet, B., and Müller-Eberhard, H. J., J. Exp. Med., in press (1978).

35. Schreiber, R. D., and Müller-Eberhard, H. J., J. Exp. Med. 140, 1324 (1974).

36. Hall, R. E., and Colten, H. R., Proc. Natl. Acad. Sci. USA 74, 1707 (1977).

37. Gorski, J. P., and Müller-Eberhard, H. J., J. Immunol. in press (1978).

38. Gorski, J. O., and Müller-Eberhard, H. J., Fed. Proc. 37, 1377 (1978).

39. Jensen, J. A., Science 155, 1122 (1967).
40. Shin, H. S., Snyderman, R., Friedman, E., Mellors, A., and Mayer, M. M., Science 162, 361 (1968).
41. Kolb, W. P., and Müller-Eberhard, H. J., J. Exp. Med. 138, 438 (1973).
42. Kolb, W. P., and Muller-Eberhard, H. J., J. Exp. Med. 143, 1131 (1976).
43. Hadding, U., Hamuro, J., and Bitter-Suermann, D., J. Immunol., in press (1978).
44. Götze, O., Bianco, C., and Cohn, Z. A., J. Immunol., in press (1978).
45. Lesavre, P., and Müller-Eberhard, H. J., Fed. Proc. 37, 1378 (1978).
46. Vogt, W., Schmidt, G., Dieminger, L., and Lynen, R., Z. Immun.-Forsch. 149, 440 (1975).
47. Davis, A. E., Rosenberg, R. D., Fenton, J. W., Bing, D. H., Rosen, F. S., and Alper, C. A., J. Immunol., in press (1978).
48. Minta, J. O., and Lepow, I. H., Immunochemistry 11, 361 (1974).
49. Götze, O., Medicus, R. G., and Müller-Eberhard, H. J., J. Immunol. 118, 525 (1977).
50. Medicus, R. G., Götze, O., and Müller-Eberhard, H. J., J. Exp. Med. 144, 1076 (1976).
51. Medicus, R. G., and Müller-Eberhard, H. J., J. Immunol., in press (1978).
52. Weiler, J. M., Daha, M. R., Austen, K. F., and Fearon, D. T., Proc. Natl. Acad. Sci. USA 73, 3268 (1976).
53. Pangburn, M. K., and Müller-Eberhard, H. J., Proc. Natl. Acad. Sci. USA 75, in press (1978).
54. Ruddy, S., Carlo, J., and Conrad, D., J. Immunol. 120, 1794 (1978).
55. Podack, E. R., Curd, J. G., Griffin, J. H., and Müller-Eberhard, H. J., Clin. Res. 26, 507A (1978).
56. Curd, J. G., Podack, E. R., Sundsmo, J. S., Griffin, J. H., and Müller-Eberhad, H. J., Clin. Res. 26, 502A (1978).
57. Fu, S. M., Kunkel, H. G., Brusman, H. P., Allen, F. H., Jr., and Fotino, M., J. Exp. Med. 140, 1108 (1974).
58. Allen, F. H., Jr., Vox Sang, 27, 382 (1974).
59. Raum, D., Glass, D., Carpenter, C., Alper, C. A.,

and Schur, P. H., J. Clin. Invest. 58, 1240 (1976).

60. Fernandez, H. N., and Hugli, T. E., J. Biol. Chem.
    252, 1826 (1977).

61. Colten, H. R., Adv. Immunol. 22, 67 (1976).

62. Loos, M., Müller, W., and Storz, R., J. Immunol.,
    in press (1978).

63. Brade, V., Fries, W., and Bentley, C., J. Immunol.,
    in press (1978).

64. Sundsmo, J. S., Curd, J. G., Kolb, W. P., and
    Müller-Eberhard, H. J., J. Immunol. 120, 855
    (1978).

65. Naff, G. B., Pensky, J., and Lepow, I. H., J. Exp.
    Med. 119, 593 (1964).

66. Valet, G., and Cooper, N. R., J. Immunol. 112,
    1667 (1974).

67. Müller-Eberhard, H. J., Polley, M. J., and Calcott,
    M. A., J. Exp. Med. 125, 359 (1967).

68. Kolb, W. P., Haxby, J. A., Arroyave, C. M., and
    Müller-Eberhard, H. J., J. Exp. Med. 138, 428
    (1973).

69. Chapitis, J., and Lepow, I. H., J. Exp. Med. 143,
    241 (1976).

70. Schreiber, R. D., Medicus, R. G., Götze, O., and
    Müller-Eberhard, H. J., J. Exp. Med. 142, 760
    (1975).

71. Vogt, W., Dames, W., Schmidt, G., and Dieminger,
    L., Immunochemistry 14, 201 (1977).

72. Cooper, N. R., Jensen, F. C., Welsh, R. M., and
    Oldstone, M. B. A., J. Exp. Med. 144, 970 (1976).

73. Bartholomew, R. M., Esser, A. F., and Müller-
    Eberhard, H. J., J. Exp. Med. 147, 844 (1978).

74. Mills, B. J., and Cooper, N. R., J. Immunol.,
    in press (1978).

75. Cooper, N. R., and Morrison, D. C., J. Immunol.,
    in press (1978).

76. Budzko, D. B., Lachmann, P. J., and McConnell, I.,
    Cell Immunol. 22, 98 (1976).

77. Platts-Mills, T. A. E., and Ishizaka, K., J.
    Immunol. 113, 348 (1974).

78. Sandberg, A. L., Oliveira, B., and Osler, A. G.,
    J. Immunol. 106, 282 (1971).

79. Joseph, B. S., Cooper, N. R., and Oldstone, M. B.
    A., J. Exp. Med. 141, 761 (1975).

80. Gigli, I., and Nelson, R. A., Jr., Exp. Cell Res. 51, 45 (1968).
81. Huber, H., Polley, M. J., Linscott, W. D., Fudenberg, H. H., and Müller-Eberhard, H. J., Science 162, 1281 (1968).
82. Müller-Eberhard, H. J., and Lepow, I. H., J. Exp. Med. 121, 816 (1965).
83. Müller-Eberhard, H. J., Dalmasso, A. P., and Calcott, M. A., J. Exp. Med. 123, 33 (1966).
84. Mayer, M. M., in "Experimental Immunochemistry" (E. A. Kabat, and M. M. Mayer, Eds.), p. 133, Charles C. Thomas, Springfield, Illinois, 1961.
85. Mayer, M. M., in "Immunochemical Approaches to Problems in Microbiology" (M. Heidelberger, and O. J. Plescia, Eds.), p. 268. Rutgers University Press, New Brunswick, New Jersey, 1961.
86. Pillemer, L., Ecker, E. E., Oncley, J. L., and Cohn, E. J., J. Exp. Med. 74, 297 (1941).
87. Becker, E. L., J. Immunol. 77, 462 (1956).
88. Lepow, I. H., Ratnoff, O. D., Rosen, F. S., and Pillemer, L., Proc. Soc. Exp. Biol. Med. 92, 32 (1956).
89. Nelson, R. A., Jr., Jensen, J., Gigli, I., and Tamura, N., Immunochemistry 3, 111 (1966).
90. Ziccardi, R. J., and Cooper, N. R., J. Immunol. 118, 2047 (1977).
91. Gigli, I., Porter, R. R., and Sim, R. B., Biochem. J. 157, 541 (1976).
92. Sim, R. B., Porter, R. R., Reid, K. B. M., and Gigli, I., Biochem. J. 163, 219 (1977).
93. Shelton, E., Yonemasu, K., and Stroud, R. M., Proc. Natl. Acad. Sci. USA 69, 65 (1972).
94. Schumaker, V. N., Calcott, M. A., Spiegelberg, H. L., and Müller-Eberhard, H. J., Biochemistry 15, 5175 (1976).
95. Cooper, N. R., Biochemistry 14, 4245 (1975).
96. Hugli, T. E., J. Biol. Chem. 250, 8293 (1975).
97. Vogt, W., Schmidt, G., v. Bufflar, B., and Dieminger, L., Immunology 34, 29 (1978).
98. Fernandez, H. N., and Hugli, T. E., J. Biol. Chem., in press (1978).
99. Pillemer, L., Blum, L., Lepow, I. H., Ross, O. A., Todd, E. W., and Wardlaw, A. C., Science 120, 279 (1954).

100. Pensky, J., Hinz, C. F., Jr., Todd, E. W., Wedg-
     wood, R. J., Boyer, J. T., and Lepow, I. H., J.
     Immunol. 100, 142 (1968).
101. Müller-Eberhard, H. J., and Götze, O., J. Exp.
     Med. 135, 1003 (1972).
102. Müller-Eberhard, H. J., Nilsson, U. R., and Arons-
     son, T., J. Exp. Med. 111, 201 (1960).
103. Götze, O., and Müller-Eberhard, H. J., J. Exp.
     Med. 134, 90s (1971).
104. Goodkofsky, I., and Lepow, I. H., J. Immunol. 107,
     1200 (1971).
105. Müller-Eberhard, H. J., Nilsson, U. R., Dalmasso,
     A. P., Polley, M. J., and Calcott, M. A., Arch.
     Path. 82, 205 (1966).
106. Müller-Eberhard, H. J., and Fjellström, K. E., J.
     Immunol. 107, 1666 (1971).
107. Cooper, N. R., J. Exp. Med. 137, 451 (1973).
108. Hunsicker, L. G., Ruddy, S., and Austen, K. F., J.
     Immunol. 110, 128 (1973).
109. Vogt, W., Dieminger, L., Lynen, R., and Schmidt,
     G., Z. Physiol. Chem. 355, 171 (1974).
110. Fearon, D. T., Austen, K. F., and Ruddy, S., J.
     Exp. Med. 138, 1305 (1973).
111. Ziegler, J. B., Alper, C. A., Rosen, F. S., Lach-
     mann, P. J., and Sherington, L., J. Clin. Invest.
     55, 668 (1975).
112. Schreiber, R. D., Götze, O., and Müller-Eberhard,
     H. J., J. Exp. Med. 144, 1062 (1976).
113. Lepow, I. H., in "Immunochemical Approaches to
     Problems in Microbiology" (M. Heidelberger, and
     O. J. Plescia, Eds.), p. 280. Rutgers University
     Press, New Brunswick, New Jersey, 1961.
114. Fearon, D. T., and Austen, K. F., Proc. Natl. Acad.
     Sci. USA 74, 1683 (1977).
115. Fearon, D. T., and Austen, K. F., J. Exp. Med. 146,
     22 (1977).
116. Lachmann, P. J., and Halbwachs, L., Clin. Exp.
     Immunol. 21, 109 (1975).
117. Schreiber, R. D., Pangburn, M. K., Lesavre, P.,
     and Müller-Eberhard, H. J., Proc. Natl. Acad. Sci.
     USA, in press (1978).
118. Kolb, W. P., Podack, E. R., and Müller-Eberhard,
     H. J., J. Immunol., in press (1978).

119. Kolb, W. P., Haxby, J. A., Arroyave, C. M., and Müller-Eberhard, H. J., J. Exp. Med. 135, 549 (1972).
120. Mayer, M. M., Scientific American 54, 229 (1973).
121. Götze, O., and Müller-Eberhard, H. J., J. Exp. Med. 132, 898 (1970).
122. Thompson, R. A., and Lachmann, P. J., J. Exp. Med. 131, 629 (1970).
123. Lachmann, P. J., and Thompson, R. A., J. Exp. Med. 131, 643 (1970).
124. Arroyave, C. M., and Müller-Eberhard, H. J., J. Immunol. 111, 536 (1973).
125. Goldlust, M. B., Shin, H. S., Hammer, C. H., and Mayer, M. M., J. Immunol. 113, 998 (1974).
126. Podack, E. R., Kolb, W. P., and Müller-Eberhard, H. J., J. Immunol., in press (1978).
127. Podack, E. R., Biesecker, G., Kolb, W. P., and Müller-Eberhard, H. J., J. Immunol., in press (1978).
128. Stolfi, R. L., J. Immunol. 100, 46 (1968).
129. Tamura, N., Shimada, A., and Chang, S., Immunology 22, 131 (1972).
130. Kolb, W. P., and Müller-Eberhard, H. J., J. Exp. Med. 141, 724 (1975).
131. McLeod, B., Baker, P., and Gewurz, H., Immunology 28, 133 (1975).
132. Podack, E. R., Kolb, W. P., and Müller-Eberhard, H. J., J. Immunol. 119, 2024 (1977).
133. Inoue, K., and Kinsky, S. C., Biochemistry, 9, 4767 (1970).
134. Hammer, C. H., Nicholson, A., and Mayer, M. M., Proc. Natl. Acad. Sci. USA 72, 5076 (1975).
135. Hammer, C. H., Shin, M. L., Abramovitz, A. S., and Mayer, M. M., J. Immunol. 119, 1 (1977).
136. Esser, A. F., and Müller-Eberhard, H. J., Abstract, VIIth Intl. Conference on Magnetic Resonance in Biological Systems, Quebec, Canada, 1976.
137. Haxby, J. A., Götze, O., Müller-Eberhard, HxxJ., and Kinsky, S. C., Proc. Natl. Acad. Sci. USA 64, 290 (1969).
138. Michaels, D. W., Abramovitz, A. S., Hammer, C. H., and Mayer, M. M., Proc. Natl. Acad. Sci. USA 73, 2852 (1976).

139. Podack, E. R., and Müller-Eberhard, H. J., J. Immunol., in press (1978).

140. Podack, E. R., Halverson, C., Esser, A. F., Kolb, W. P., and Muller-Eberhard, H. J., J. Immunol., in press (1978).

141. Griffin, F. M., Bianco, C., and Silverstein, S. C., J. Exp. Med. 141, 1269 (1975).

142. Bianco, C., Griffin, F. M., and Silverstein, S. C., J. Exp. Med. 141, 1278 (1975).

143. West, C. D., Davis, N. C., Forristal, J., Herbert, J., and Spitzer, R., J. Immunol. 96, 650 (1966).

144. Bokisch, V. A., Müller-Eberhard, H. J., and Cochrane, C. G., J. Exp. Med. 129, 1109 (1969).

145. Ruddy, S., and Austen, K. F., J. Immunol. 107, 742 (1971).

146. Bokisch, V. A., Dierich, M. P., and Müller-Eberhard, H. J., Proc. Natl. Acad. Sci. USA 72 1989 (1975).

147. Ross, G. D., Polley, M. J., Rabellion, E. M., and Grey, H. M., J. Exp. Med., 138, 798 (1973).

148. Ghebrehiwet, B., and Müller-Eberhard, H. J., J. Immunol., in press (1978).

149. Cochrane, C. G., Muller-Eberhard, H. J., and Aikin, B. S., J. Immunol. 105, 55 (1970).

150. Henson, P. M., and Cochrane, C. G., J. Exp. Med. 133, 554 (1971).

151. Lennon, V. A., Seybold, M. E., Lindstrom, J. M., Cochrane, C. G., and Ulevitch, R., J. Exp. Med. 147, 973 (1978).

152. Hill, J. H., and Ward, P. A., J. Exp. Med. 133, 885 (1971).

153. Maroko, P. R., Carpenter, C. B., Chiariello, M., Fishbein, M. C., Radvany, P., Knostman, J. D., and Hale, S. L., J. Clin. Invest. 61, 661 (1978).

154. O'Rourke, R. A., Crawford, M. H., Ghidoni, J. J., Grover, F. L., Olson, M. S., and Pinkard, R. N., Clin. Res. 26, 484A (1978).

155. Alper, C. A., and Balavitch, D., Science 191, 1275 (1976).

156. Gigli, I., in "Comprehensive Immunology. Biological Amplification Systems in Immunology" (N. K. Day, and R. A. Good, Eds.), p. 295, Vol. 2. Plenum Publishing Co., New York, 1977.

157. Ruddy, S., Carpenter, C. B., Chin, K. W., Knost-
     man, J. N., Soter, N. A., Götze, O., Müller-
     Eberhard, H. J., and Austen, K. F., Medicine 54,
     165 (1975).
158. Hunsicker, L. G., Ruddy, S., Carpenter, C. B.,
     Schur, P. H., Merrill, J. P., Müller-Eberhard,
     H. J., and Austen, K. F., N. Engl. J. Med. 287,
     835 (1972).
159. Sissons, J. G. P., Liebowitch, J., Amos, N., and
     Peters, D. K., J. Clin. Invest. 59, 704 (1977).
160. Lepow, I. H., Willms-Kretschmer, K., Patrick, R.
     A., and Rosen, F. S., Amer. J. Path. 61, 13
     (1970).
161. Wuepper, K. D., Bokisch, V. A., and Müller-
     Eberhard, H. J., and Stoughton, R. B., Clin.
     Exp. Immunol. 11, 13 (1972).
162. Hugli, T. E., Vallota, E. H., and Müller-Eberhard,
     H. J., J. Biol. Chem. 250, 1472 (1975).
163. Fernandez, H. N., and Hugli, T. E., J. Immunol.
     117, 1688 (1976).
164. Hugli, T. E., Morgan, W. T., and Müller-Eberhard,
     H. J., J. Biol. Chem. 250, 1479 (1975).
165. Morgan, W. T., Vallota, E. H., and Müller-Eberhard,
     H. J., Biochem. Biophys. Res. Comm. 57, 572
     (1974).
166. Friedberger, E., Z. Immun.-Forsch., 2, 208 (1909).
167. Stegemann, H., Vogt, W., and Friedberg, K. D.,
     Z. Physiol. Chem., 337, 269 (1964).
168. Vogt, W., and Schmidt, G., Experientia, 20, 207
     (1964).
169. Liefländer, M., Dielenberg, D., Schmidt, G., and
     Vogt, W., Z. Physiol. Chem. 353, 385 (1972).
170. Budzko, D. B., Bokisch, V. A., and Müller-Eberhard,
     H. J., Biochemistry 10, 1166 (1971).
171. Hugli, T. E., and Erickson, B. W., Proc. Natl.
     Acad. Sci. USA 74, 1826 (1977).

172. Rother, K., _Europ. J. Immunol._ 2, 550 (1972).
173. Alper, C. A., Colten, H. R., Rosen, F. S.,
      Rabson, A. R., Macnab, G. M., and Gear, J. S. S.,
      Lancet ii, 1179 (1972).
174. Rosen, F. S., and Lachmann, P. J., in "Immune De-
      ficiency. Springer Seminars in Immunopathology"
      (P.A. Miescher, and H. J. Müller-Eberhard, Eds.,
      M. D. Cooper, Guest Ed.), Vol. 1. Springer
      Verlag, New York, in press, 1978.

THE EARLY PHASE OF BLOOD COAGULATION

Earl W. Davie, Kazuo Fujikawa, Walter Kisiel,
Kotoku Kurachi, and Ronald L. Heimark

University of Washington

*Factor XII (Hageman factor) is a plasma pro-
tein that participates in the early phase of the
intrinsic and extrinsic pathways of blood coagulation. Bovine factor XII is a single-chain
molecule which is converted to factor XII_a by
minor proteolysis. Factor XII_a is composed of
a heavy and a light chain held together by a
disulfide bond(s). The active site in this
protein is in the light chain of the molecule.
In the intrinsic pathway, factor XII_a converts
factor XI to factor XI_a in the presence of high
molecular weight kininogen. In the extrinsic pathway, factor XII_a converts factor VII to factor
VII_a.*

When plasma is placed in a glass test tube, a number of different physiological reactions are initiated
including blood coagulation (1-3), fibrinolysis (4-6),
and kinin formation (7). Several proteins including
factor XII (Hageman factor), prekallikrein (Fletcher
factor), and high molecular weight kininogen (HMW
kininogen) participate in these reactions which are
triggered by the presence of a negatively charged surface such as kaolin or glass.
Present evidence indicates that the three proteins
above are involved in both the intrinsic and extrinsic
pathways of blood coagulation. In the intrinsic pathway, factor XII, prekallikrein, and HMW kininogen

115

interact in the presence of a surface leading to the
activation of factor XI (8-11). These reactions are
shown in Figure 1.

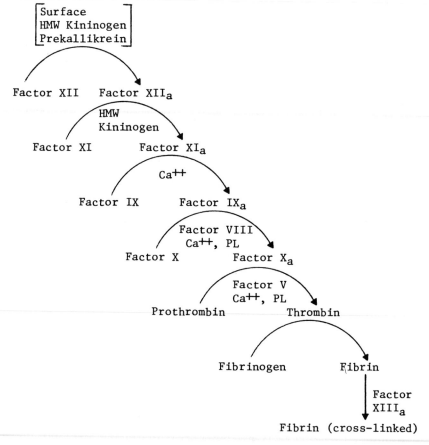

Fig. 1. Abbreviated mechanism for the surface
activation of blood coagulation in the intrinsic pathway. PL refers to phospholipid. Modified from Davie
et al. (12).

Surface activation may also be of major importance in the extrinsic pathway leading to the activation of factor VII (proconvertin) (3,13-20). Indeed, it has been shown recently in our laboratory that factor XII$_a$ will activate factor VII directly (21). Factor VIIa in the presence of tissue factor will then activate factor X (Stuart factor) which in turn will activate prothrombin in the presence of factor V (proaccelerin), calcium, and phospholipid. These reactions are shown in Figure 2.

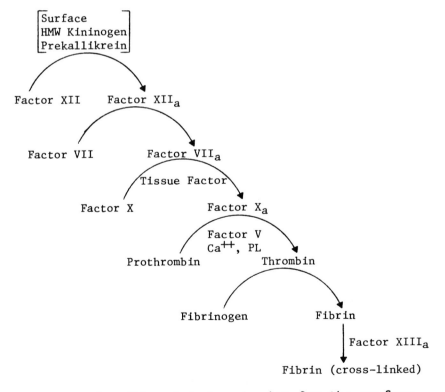

Fig. 2. Abbreviated mechanism for the surface activation of blood coagulation in the extrinsic pathway. PL refers to phospholipid.

This pathway bypasses factor XI (plasma thromboplastin antecedent), factor IX (Christmas factor), and factor VIII (antihemophilic factor).

In recent years, many of the proteins participating in the early phase of blood coagulation have been isolated and characterized. In our laboratory, we have focused our attention on developing methods for the isolation of factor XII, prekallikrein, HMW kininogen, factor XI, and factor VII from bovine sources. Bovine blood was employed as a starting material since these coagulation factors are trace proteins and a large amount of starting material was necessary. Also, substantial quantities of highly purified preparations are necessary in order to characterize these proteins and study their interaction on a molecular basis. Recently, we have isolated factor XII in a highly purified form and have been able to prepare approximately 15 mg of protein from 15 liters of plasma (22). The overall yield was about 20%. The isolation procedure involves heparin-agarose, DEAE-Sephadex, CM-cellulose, arginine-agarose, and benzamidine-agarose column chromatography. Bovine factor XII is a single-chain glycoprotein with a molecular weight of 74,000 as determined by sedimentation equilibrium. The carbohydrate present included 3.4% hexose, 4.7% N-acetyl-hexosamine, and 5.4% N-acetylneuraminic acid.

The amino-terminal sequence of bovine factor XII was found to be: Thr-Pro-Pro-Trp-Lys-Gly-Pro-Lys-Lys-His-. It is of interest to note that this sequence is about 50% homologous with the active site region of a number of protease inhibitors(23). Whether this amino-terminal region partially masks the active site in factor XII or has inhibitory activity toward some other serine protease(s) is not known.

Factor XII has also been isolated in a two-chain form from bovine plasma (24). This protein was identified as factor XII$_a$. It has a specific activity of 1.5 to 2.0 fold greater than factor XII in the coagulant assay. It is composed of a heavy and a light chain, and these two chains are held together by a disulfide bond(s). The heavy chain originates from the amino-terminal region of the precursor protein and has a molecular weight of approximately 46,000. Arginine has been identified as its carboxyl-terminal residue. The light chain originates from the carboxyl-terminal region of the precursor protein and has a molecular weight of approximately 28,000. The amino-terminal sequence of the light chain is Val-Val-Gly-Gly-Leu-Val-Ala-Leu-Pro-Gly-Ala-. These data suggest that factor XII_a was formed from factor XII by the cleavage of a specific internal arginyl-valine peptide bond. The light chain of factor XII_a also contains the active site region of the enzyme, and this sequence was shown to be Asp-Ala-Cys-Gln-Gly-Asp-Ser-Gly-Gly-Pro-Leu-Val-Cys-. The amino-terminal sequence of the light chain and the active site sequence are homologous with the corresponding region of a number of plasma serine proteases as well as pancreatic trypsin.

Thus far, four protein substrates for factor XII_a have been identified in plasma. These include factor XI, factor VII, prekallikrein, and plasminogen proactivator. The activation of factor XI in the intrinsic pathway involves limited proteolysis giving rise to factor XI_a. This reaction is stimulated by HMW kininogen (10,25). The peptide bond(s) in factor XI that is cleaved during this activation reaction has not been identified. The activation of factor VII in the extrinsic pathway also involves limited proteolysis giving rise to factor VII_a (21). In this reaction, a specific arginyl-isoleucine bond in factor VII is hydrolyzed. Factor VII_a is composed of a heavy and a light chain, and these chains are held together by a disulfide bond(s). Whether or not this activation reaction is stimulated by HMW kininogen is not known. The peptide bond(s) that is split by factor XII_a in prekallikrein or plasminogen proactivator during the conversion of these proteins to enzymes is also not known.

Factor XII_a also has esterase activity toward tosyl-L-arginine methyl ester and benzoyl-L-arginine ethyl ester (24,26,27). It has little or no activity toward tosyl-L-lysine methyl ester, tosyl-L-methionine ethyl ester, or benzoyl-L-alanine methyl ester (27). Thus, the specificity of factor XII_a is directed toward peptide bonds containing arginine. This specificity is consistent with the presence of an aspartic acid six

residues prior to the active site serine (22). This
residue is located in the bottom of the binding pocket
in trypsin (28) and forms an ion pair with a basic
amino acid residue in the substrate (29-33). Thus, it
appears probable that the aspartic acid residue
occupying the same position in factor XII_a gives this
enzyme a similar specificity toward peptide bonds containing arginine.

Factor XII_a is readily inhibited by a number of
serine protease inhibitors. The inactivation of
factor XII_a by diisopropyl phosphorofluoridate (DFP)
was first demonstrated by Becker in 1960 (34). Similar
results have been found by other investigators (24,35,
36). The DFP binding site in bovine factor XII_a is
located in the light chain of the molecule (24). This
is consistent with the data showing the presence of an
active site serine sequence in the light chain as discussed above. DFP does not inhibit bovine factor XII
in the presence or absence of kaolin (24). Human
factor XII (single chain) appears to differ from the
bovine preparation in that it has been reported to be
inhibited by DFP (37).

Factor XII_a is also sensitive to antithrombin III.
In this case, the human preparation also seems to differ from the bovine preparation. Bovine factor XII_a
is rapidly inhibited by antithrombin III, and this
reaction is accelerated by the presence of heparin
(24). The inhibitor is bound to the light chain of the
enzyme. Bovine factor XII, however, is resistant to
antithrombin III in the presence or absence of heparin.
Human factor XII (single chain), in contrast has been
reported to be inhibited by antithrombin III (37).
Furthermore, a complex between human factor XII and
antithrombin III was demonstrated by sodium dodecyl
sulfate-polyacrylamide gel electrophoresis. Thus,
human factor XII and bovine factor XII clearly differ
in their sensitivity to antithrombin III. Other serine
protease inhibitors such as soybean trypsin inhibitor,
trasylol, and ovomucoid trypsin inhibitor have little
or no effect on factor XII_a, while lima bean trypsin
inhibitor and benzamidine have been reported to inhibit
the enzyme (27,36,38).

Factor XII can also be converted to an enzyme in the absence of proteolysis (39,40). This reaction occurs on a surface such as glass or kaolin and in the presence of a suitable substrate such as factor XI or prekallikrein (25,41). Substrates such as factor VII or benzoyl-L-arginine ethyl ester are hydrolyzed very slowly or not at all by bovine factor XII in the presence of kaolin (22). Furthermore, no inhibition of bovine factor XII has been observed by DFP or antithrombin III in the presence of kaolin (22). Thus, factor XI or prekallikrein appear to induce an activation of surface-bound factor XII permitting it to hydrolyze these specific substrates. The role of HMW kininogen in these reactions is not known.

I. REFERENCES

1. Ratnoff, O. D., and Colopy, J. E., J. Clin. Invest. 34, 602 (1955).
2. Ratnoff, O. D., and Rosenblum, J. M., Amer. J. Med. 25, 160 (1958).
3. Soulier, J. P., and Prou-Wartelle, O., Brit J. Haematol. 6, 88 (1960).
4. Niewiarowski, S., and Prou-Wartelle, O., Thromb. Diath. Haemorrh. 3, 593 (1959).
5. Iatridis, S. G., and Ferguson, J. H., Thromb. Diath. Haemorrh. 6, 411 (1961).
6. Ogston, D., Ogston, C. M., Ratnoff, O. D., and Forbes, C. D., J. Clin. Invest. 48, 1786 (1969).
7. Margolis, J., Australian J. Exptl. Biol. Med. Sci. 37, 289 (1959).
8. Schiffman, S., and Lee, P., Brit. J. Haematol. 27, 101 (1974).
9. Schiffman, S., Lee, P., and Waldman, R., Thromb. Res. 6, 451 (1975).
10. Griffin, J. G., and Cochrane, C. G., Proc. Natl. Acad. Sci. U. S. A. 73, 2554 (1976).
11. Chan, J. Y., Habal, F. M., Burrowes, C. E., and Movat, H. Z., Thromb. Res. 9, 423 (1976).
12. Davie, E. W., Fujikawa, K., Kurachi, K., Koide, T., and Kisiel, W., in "Proteolysis and Physiological Regulation" (K.W. Robbins and K. Brew, Eds.),

Miami Winter Symposia, Vol. 11, p. 189, Academic Press, New York, 1976.

13. Rapaport, S. I., Aas, K., and Owren, P. A., J. Clin. Invest. 34, 9 (1955).
14. Altman, R., and Hemker, H. C., Thromb. Diath. Haemorrh. 18, 525 (1967).
15. Shanberge, J. N., and Matsuoka, T., Thromb. Diath. Haemorrh. 15, 442 (1966).
16. Gjonnaess, H., Thromb. Diath. Haemorrh. 28, 182 (1972).
17. Gjonnaess, H., Thromb. Diath. Haemorrh. 28, 194 (1972).
18. Laake, K., and Ellingsen, R., Thromb. Res. 5, 539 (1974).
19. Laake, K., and Osterud, B., Thromb. Res. 5, 759 (1974).
20. Saito, H., and Ratnoff, O. D., J. Lab. Clin. Med. 85, 405 (1975).
21. Kisiel, W., Fujikawa, K., and Davie, E. W., manuscript submitted to Biochemistry.
22. Fujikawa, K., Walsh, K. A., and Davie, E. W., Biochemistry in press (1977).
23. Laskowski, M., Jr., Kato, I., Leary, T. R., Schrode, J., and Sealock, R. W., in "Proteinase Inhibitors" (H. Fritz, H. Tschesche, L. J. Greene, and E. Truscheit, Eds.), Bayer-Symposium V, p. 597. Springer-Verlag, New York, (1974).
24. Fujikawa, K., Kurachi, K., and Davie, E. W., manuscript submitted to Biochemistry.
25. Kurachi, K., unpublished results.
26. Schoenmakers, J. G. G., Matze, R., Haanen, C., and Zilliken, F., Biochim. Biophys. Acta 101, 166 (1965).
27. Komiya, M., Nagasawa, S., and Suzuki, T., J. Biochem. 72, 1205 (1972).
28. Stroud, R. M., Kay, L. M., and Dickerson, R. E., Cold Spring Harbor Symp. Quant. Biol. 36, 125 (1971).
29. Mares-Guia, M., and Shaw, E., J. Biol. Chem. 240, 1579 (1965).
30. Ruhlmann, A., Kulka, D., Schwager, P., Bartels, K., and Huber, R., J. Mol. Biol. 77, 417 (1973).
31. Blow, D. M., Janin, J., and Sweet, R. M., Nature

(London) 249, 54 (1974).

32. Sweet, R. M., Wright, H. T., Janin, J., Chothia, C. H., and Blow, D. M., Biochemistry 13, 4214 (1974).

33. Krieger, M., Kay, L. M., and Stroud, R. M., J. Mol. Biol. 83, 209 (1974).

34. Becker, E. L., J. Lab. Clin. Med. 56, 136 (1960).

35. Schoenmakers, J. G. G., Matze, R., Haanen, C., and Zilliken, F., Biochim. Biophys. Acta 93, 433 (1964).

36. Wuepper, K. D., and Cochrane, C. G., J. Clin. Invest. 50, 100 (1972).

37. Stead, N., Kaplan, A. P., and Rotenberg, R. D., J. Biol. Chem. 251, 6481 (1976).

38. Johnston, A. R., Cochrane, C. G., and Revak, S. G., J. Immunol. 113, 103 (1974).

39. Cochrane, C. G., Revak, S. D., and Wuepper, K. D., J. Exp. Med. 138, 1564 (1973).

40. Morrison, D. C., and Cochrane, C. G., J. Exp. Med. 140, 797 (1974).

41. Heimark, R., unpublished results.

CASCADE EVENTS IN MAST CELL ACTIVATION AND FUNCTION[1]

Roger W. Yurt[2] and K. Frank Austen

*Harvard Medical School and Robert B. Brigham Hospital
Boston, Massachusetts*

The mast cell, with its specific immunologic recognition units, is positioned in tissues at portals of possible entry of noxious, exogenous materials. As it does not need to arrive at tissue sites in response to injury, it in fact may well be the sentinel cell of the effector immune responses. During activation a series of intracellular biochemical steps occur, including the uncovering of esterase activity, calcium ion-dependent events and modulation by the intracellular level of the cyclic nucleotides. The end result of mast cell activation is the generation and release of a diverse group of highly active pharmacologic materials which possess all the necessary functions for a fully expressed but controlled local inflammatory response. The activity of the mediators may be expressed directly or may require a "second activation" to fully express activity. These chemical mediators cause an early local host response and account for a later ingress of antibody and complement as well as the influx of specific phagocytic cells. The natural regulation of this process includes: the intensity of the mast cell activating event; the intracellular controls of generation and release of mediators; the action of primary and secondary mediators on the mast cell; the expression of mediator activity; and the rate at which released mediators undergo biodegradation.

[1]Supported by grants AI-07722, AI-10356, and RR-05669 from the National Institutes of Health.
[2]Postdoctoral trainee supported by training grant AM-07031 from the National Institutes of Health.

The mast cell by virtue of its specific IgE receptor and
its mucosal (1,2) and perivenular (3) locations serves as an
immunologic sentinel to the invasion of foreign substances.
Mast cells are present in the mucosal surfaces and lumen of
bronchi in rhesus monkeys and humans (1,2) and occupy intra-
epithelial positions in adenoids, tonsils and nasal polyps
(4). The presence of IgE in the mast cells in human skin and
respiratory tissue has been established by immunofluorescence
with specific anti-human IgE (4) and by generation of chemical mediators by anti-human IgE from human lung (5) and nasal
polyp fragments (6) and fractionated human lung cells (7).
Once the mast cell is perturbed, a cascade of steps including
calcium ion influx, the uncovering of a serine esterase, an
energy-requiring phase, a second calcium-dependent event and
a fall in intracellular cyclic 3',5'-adenosine monophosphate
(cyclic AMP) occurs. This cellular activation and coupled
secretory reaction culminate in the release of preformed mediators in a fully active form, such as histamine and the
eosinophilotactic tetrapeptide known as eosinophil chemotactic factor of anaphylaxis (ECF-A), and in the discharge of
granule-associated heparin and chymotrypsin-like protease
(chymase) which require further release to express full activity. Activation, in addition, leads to the generation and
release of unstored factors such as the slow reacting substance of anaphylaxis (SRS-A), platelet activating factor
(PAF) and lipid chemotactic factors. The array of mediators
may then alter the microenvironment to allow the influx of
plasma proteins, including immunoglobulin and complement components, and the ingress of peripheral blood cells. The initial or humoral phase of the local host response is mediated
by substances such as histamine and SRS-A and may be controlled locally or terminated by the ingress of cells such as
eosinophils which biodegrade mediators during the cellular
phase. During this cellular phase the host response is
greatly amplified by the influx of polymorphonuclear leukocytes that are drawn to the site by ECF-A, oligo-ECF, and
neutrophil chemotactic factor (NCF); the combined action of
cells, antibody, and complement provides local physiologic
defense against organisms as diverse as pyogenic bacteria and
schistosomula.
     Clinically, the uncontrolled mast cell humoral response
may be seen as urticaria, angioedema, exacerbations of rhini-
tis or asthma, and systemic anaphylaxis. The cellular response may progress to an inflammatory reaction as seen in
histologic studies of nasal polyps and bronchial tissue of
atopic allergic individuals (Fig. 1).

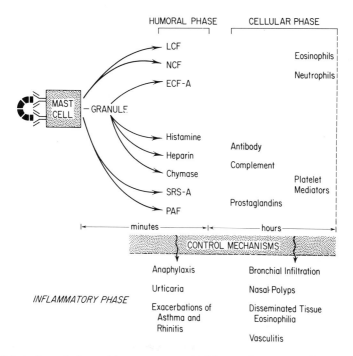

Fig. 1. Schematic representation of mast cell-dependent humoral and cellular responses.

Both phases of the mast cell response are manifested in the cutaneous reaction to antigenic challenge of sensitized individuals (8-10), which is biphasic in presenting an edematous urticarial phase followed by an indurated inflammatory phase.

I. MAST CELL ACTIVATION AND DEGRANULATION

A. Biochemical Requirements

This discussion will center primarily on immunologic activation of dispersed cells and tissues but will refer to other activators when pertinent. Human basophil activation by antigen (11,12) and rat mast cell activation by antibody (13,14) require that the stimulus be bivalent or multivalent, indicating that ligand bridging of the surface-bound IgE is necessary for the initial phase of membrane perturbation (Fig. 2).

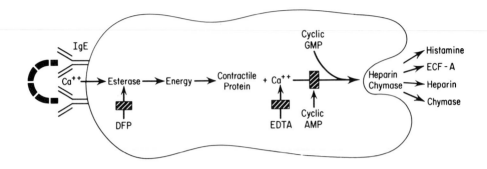

Fig. 2. Schematic representation of the IgE-dependent biochemical events in mast cell activation and secretion of preformed mediators.

Extensive clustering or capping of IgE is not required for activation of the basophil (12,15) or the isolated rat peritoneal mast cell (14) where the maximum size of a stimulatory cluster is less than 10 immunoglobulin molecules. Soon after antigen challenge of human lung fragments (16), an extracellular calcium ion-dependent diisopropylfluorophosphate (DFP)-sensitive serine esterase is uncovered. A rat mast cell enzyme uncovered during IgE-dependent cell activation appears to be chymotrypsin-like in its inhibition profile (17,18), and even release of histamine by rabbit neutrophil cationic protein requires a DFP-inhibitable esterase (19). If the irreversible inhibitor is removed before membrane perturbation by antigen (16) or cationic protein (19), no inhibitory effect is seen, thus demonstrating that the esterase in the resting cell is protected. Indirect support for the action of a chymotrypsin-like esterase is derived from reports that α-chymotrypsin induces a noncytolytic, calcium ion, and adenosine triphosphate (ATP)-dependent (20-22) histamine release from isolated rat mast cells that is similar by ultrastructural criteria to antigen-induced release (22,23); further, the active site of the enzyme was required for this effect. The fact that high concentrations of DFP do not inhibit the release of histamine induced by polymyxin B and 48/80 (22,24) suggests that these agents by-pass the early activation step. That trypsin-treated rat mast cells respond to the polyamines but not to α-chymotrypsin supports the concept of a by-pass of early steps required during physiologic activation (22). In contrast to histamine release, inhibition of SRS-A release by DFP is not reversed by washing sensitized human lung fragments

(16) before antigen activation; this finding may be analogous
to those in chemotaxis (25) and in phagocytosis (26), where
both an "activated" and "activatable" esterase have been
implicated. Whether the granule-associated chymotrypsin
represents the serine esterase that is essential to mast cell
activation remains to be determined.

An energy-requiring step is necessary to maintain the
sequence leading to noncytotoxic mast cell mediator release
in all activation systems studied (27,28). When reactions
are conducted in the presence of 2-deoxyglucose (2-DG) in the
absence of glucose, release of mediators by anti-F(ab')$_2$
challenge of isolated rat mast cells (27) or antigen challenge of human lung (16) is suppressed. This ability of 2-DG
to prevent release is attributed to a requirement for glyco-
lysis and kinetically occurs after the uncovering of the
esterase in human lung fragments (16). Oxidative phosphorylation is an alternative energy source, since histamine release
is ablated by incubation of mast cells in antimycin A (27).
A second calcium-dependent, EDTA-inhibitable step in the
activation cascade of human lung fragments occurs after and
separate from the early extracellular calcium ion-dependent
step and after the energy-dependent step. A requirement for
intracellular calcium is compatible with the activation of
contractile protein that is inhibited through phosphorylation
by a cyclic AMP-dependent kinase (29). Recent ultrastructural
studies of antigen-stimulated rat mast cells do indicate the
presence of contractile filaments in degranulating cells (30).
This fact, coupled with the demonstrated augmentation and inhibition of the basophil release reaction by microtubular active agents D_2O and colchicine, respectively (31), supports
the theory of a cyclic nucleotide-regulated secretory process.

Under conditions that elevate cellular cyclic AMP levels,
such as stimulation with the beta-adrenergic agonist isopro-
terenol, prostaglandin E_1 (PGE_1) or cholera toxin, mediator
release is inhibited in human lung fragments (16). This
effect is also seen in studies of isolated rat mast cells
where a synergistic increase in cyclic AMP secondary to introduction of a phosphodiesterase inhibitor with PGE_1 produced a synergistic inhibition of mediator release (27).
Conversely, enhancement of immunologic release of mediators
is observed in human lung fragments when cyclic AMP levels
are depressed by stimulation with the alpha adrenergic agents
norepinephrine or phenylepinephrine in the presence or absence of the β-adrenergic antagonist propranolol (32). In
addition, cholinomimetic agents enhance the immunologic release of mediators in association with an increase in tissue
levels of cyclic 3',5'-guanosine monophosphate (cyclic GMP)

but not of cyclic AMP (32). The role of cyclic nucleotides
in mast cell activation is complex (29) but most evidence suggests that a critical early fall in cyclic AMP is involved.

B. Degranulation

Recent ultrastructural data (33) show that the earliest
morphologic correlate of activation of rat mast cells by
ferritin-coupled anti-immunoglobulin or concanavalin A consists of a ruffling of the granule membrane and the development of a halo between it and the granule; subsequently, the
granule and plasma membrane fuse to form a pentilaminar
structure. The ferritin-labeled material becomes concentrated at the periphery of the fused membrane; and a lipid-rich,
protein-deficient bleb forms over the fusion area. The bleb
is then "pinched off" the membrane to expose the granule to
or allow its egress into the extracellular fluid. Freeze-
fracture electron microscopy of rat mast cells activated by
polymyxin B confirms both bleb formation and the concentration
of intramembranous particles to the periphery of the bleb (34).
These findings are consistent with earlier morphologic reports
that rat mast cells challenged with 48/80 (30) or antigen (23)
contain granules exposed to extracellular fluid by virtue of
continuity of the plasma and granule membranes.
The chemical manifestation of degranulation correlates
directly with the morphologic data and appears to consist of
two distinct mechanisms for release of functional mediators
to the microenvironment. The initial exposure of the granule
to the ion-containing extracellular fluid releases those mediators bound by low affinity charge interaction, such as histamine (35,36) and ECF-A (37); this mechanism does not require
that the entire granule matrix be mechanically separated from
the cell. However, studies with activation by 48/80 (38,39)
indicate that heparin, a component of the residual matrix
(40-42), is released. Furthermore, the finding that rat mast
cells challenged with anti-F(ab')$_2$ release granule-bound
heparin in a linear relationship to histamine (43) confirms
that physiologic degranulation (43) involves both ion exchange
of mediators as well as actual extrusion of the granule matrix
from the confines of the cell.
The fate of the extruded granule matrix and its major
components, heparin and the chymotrypsin-like enzyme desig-
nated chymase (44), is not well defined. There is evidence
to suggest that it may be phagocytosed, but solubilization
under physiologic conditions has not been shown. Injection of
isolated, radiolabeled rat peritoneal mast cell granules into
rat skin leads to their uptake by fibroblasts (45); and rat
eosinophils (46,47), macrophages and fibroblasts (48) have

been observed to contain metachromatic granules after intraperitoneal mast cell degranulation induced by 48/80 (46-48)
and distilled water (48). The result of such phagocytic
processes is not known although murine macrophages apparently
contain enzymes capable of cleaving ^{35}S from radiolabeled
commercial heparin (49). The very low isoelectric point of
native rat mast cell heparin (40), which migrates toward the
anode at pH 3.5 during glycerol gradient electrophoresis (Yurt
and Austen, unpublished observation), and the high isoelectric
point of the chymase, apparently $\cong$ 9 (40,50), create a matrix
that only solubilizes under extremes of salt concentration,
e.g., greater than 1 M NaCl (50), 1-2 M KCl (42), or pH 10-11
(40). A recent intramolecular approach to solubilization of
the granules that may be of physiologic interest and the
functional consequences of such events will be discussed with
regard to heparin and chymase.

II. PHYSICOCHEMICAL AND FUNCTIONAL CHARACTERISTICS OF PRE-
     FORMED MEDIATORS

     The mast cell mediators are at varying stages of chemical
defintion (Table 1).

TABLE I. CHEMICAL MEDIATORS OF IMMEDIATE HYPERSENSITIVITY

Category	Mediator	Structural Characteristics	Assay(s)	Other Functions	Inactivation
I. PREFORMED	Histamine	β-imidazolyl-ethylamine MW = 111	Contraction of guinea pig ileum Specific radio-labeling by histamine N-methyl-transferase	Increased vascular permeability Elevation of cyclic AMP Enhancement of eosinophil migration	Histaminase: Histamine methyl-transferase
	Eosinophil chemotactic factor of anaphylaxis (ECF-A)	Hydrophobic acidic tetra-peptides MW = 360-390	Chemotactic attraction of eosinophils	Eosinophil deactivation	Aminopeptidase; carboxy-peptidase A
	Neutrophil chemotactic factor (NCF)	Protein MW > 160,000	Chemotactic attraction of neutrophils	Neutrophil deactivation	--
	Heparin	Macromolecular acidic proteo-glycan MW ≈ 750,000	Anti-thrombin activation or metachromasia	Anticoagulation	--
	Chymase	MW = 25,000-29,000	Hydrolysis of BTEE	Proteolysis	--
	N-acetyl-β-glucosamini-dase	Protein MW = 150,000	Hydrolysis of N-acetyl-β-glucosamini-dase parani-troaniline	--	--
II. UNSTORED	Slow reacting substance of anaphylaxis (SRS-A)	Acidic hydro-philic sulfate ester MW = 400	Contraction of antihistamine-treated guinea pig ileum	Contraction of human bronchiole Increased vascu-lar permeability	Arylsulfatase B
	Platelet acti-vating factors (PAF)	Lipid-like MW ≈ 300-500	Release of ^{14}C-5HT from platelets	Aggregation	Phospholipase D (for rat PAF)
	Lipid chemo-tactic factor (LCF)	Lipid-like	Chemotactic attraction of neutrophils	Neutrophil deactivation	--

Histamine has been structurally and functionally characterized
(51-53) and will not be further discussed in detail here.
The other mediators will be reviewed under the classification
of preformed (extractable from tissue or cells) or generated
(requiring activation of the cell).

A. Heparin

    The basophilic granules which give a metachromatic color
were the basis for the definition of the mast cell by Ehrlich
(54) nearly a century ago. The highly sulfated glycosaminoglycan heparin has been shown to account for this metachromasia of the rat peritoneal mast cell (55-57). Heparin has
been isolated and partially characterized in mast cell-rich
tissues such as bovine (58) and ox (59) liver capsules, rat
skin (60), murine mastocytoma (61), and lung of several
species including beef, pork, sheep (62) and human (63). In
contrast the basophilic granules of the guinea pig basophil
appear to contain primarily chondroitin and dermatan sulfate
rather than heparin based on characterization of the glycosaminoglycan chains obtained after cell disruption and proteolytic digestion of granules (64).
    Heparin has been released from rat mast cells in a dose-
dependent manner by 48/80 as measured by release of ^{35}S
(38,65) label or of metachromatic material (39). Furthermore,
purified rat peritoneal mast cells labeled in vivo or in vitro
with ^{35}S and challenged with anti-F(ab')$_2$, anti-IgE or the
calcium ionophore release ^{35}S-labeled heparin as defined by
metachromasia after extraction of the granule with 1 M NaCl
and 0.05 M NaOH and isolation on Dowex 1 (43). The released
and residual heparin as well as that extracted from unchal-
lenged cells has 10-20% of the anticoagulant and antithrombin
cofactor activity of commercial heparin and exhibits an
apparent molecular weight of 750,000 by gel filtration. Based
on the incorporation of ^{3}H and ^{35}S into the heparin molecule
during in vitro incubation of purified rat mast cells with
^{3}H-serine and ^{35}S and the degradation, termed β elimination,
by alkali, to a molecular weight of 40,000 with the concurrent release of ^{3}H peptides, this macromolecule appears to be
a proteoglycan. However, as it is not susceptible to proteolytic digestion (56), the peptide core must be sterically
inaccessible due to the density and nature of the glycosaminoglycan side chains.
    To investigate further the mechanisms of cleavage of
macromolecular heparin, purified rat peritoneal mast cells
were incubated for 4 hours with ^{35}S and extracted with 1 M
NaCl; the ^{35}S-heparin was isolated in the 3 M eluate from
Dowex 1 (56) and treated with oxidizing or reducing agents.
Portions of the labeled heparin were incubated at 37°C in 0.3

ml 0.01 M potassium phosphate buffer, pH 7.4 with a 0.4 M
NaCl solution containing 4.4 μm CuSO₄, 2.4 mM ascorbate,
and 2.4 mM ascorbate and 4.4 μM CuSO₄, respectively, or in
buffer alone for varying periods of time. The reaction mix-
tures were filtered on a 1 x 50 cm column of Sepharose 4B
together with internal standards of blue dextran, chondroitin 6-sulfate (C-6-S), commercial heparin and phenol red.
The ^{35}S-heparin in buffer or treated with CuSO₄ filtered with
an apparent molecular weight of 750,000 (Fig. 3a).

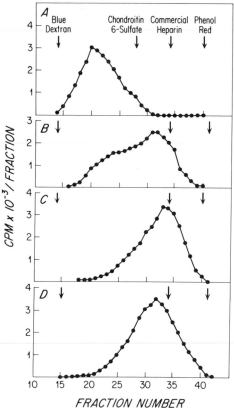

Fig. 3. Gel filtration on Sepharose 4B of ^{35}S-heparin
from rat mast cells treated with buffer, ascorbate, or
ascorbate and cupric sulfate. Portions of ^{35}S-heparin were
treated with buffer (A), ascorbate for 1.5 (B) or 6 hours
(C), or ascorbate with cupric sulfate for 1 hour (D) and
filtered together with internal standards of blue dextran,
commercial heparin (average molecular weight = 12,000) and
phenol red. The positions of the internal standards and
chondroitin 6-sulfate (approximate molecular weight -
60,000), which was filtered separately on the same column are
shown. Fractions were assayed for ^{35}S (●——●).

Approximately 50% of the ascorbate-treated heparin showed a decrease in size to an approximate molecular weight of 20,000 after 1.5 hours of incubation (Fig. 3b), and most of the heparin was degraded to this size after 6 hours of incubation (Fig. 3c). In the presence of 4.4 μM CuSO$_4$ and 2.4 mM ascorbate, the ^{35}S-labeled heparin was degraded to the low molecular weight form (Fig. 3d) after only 1 hour. In additional experiments in which ^{35}S-heparin was incubated for 2 hours at 37°C with the same concentration of ascorbate and copper in a nitrogen atmosphere or in a 0.1 M concentration of the reducing agent dithiothreitol (DTT), degradation was prevented. Treatment of the heparin with an oxidizing agent, 4 mM potassium permanganate, also caused a reduction in the molecular weight of the ^{35}S-heparin to approximately 20,000. These findings are consistent with the observed decrease in viscosity of rat skin macromolecular heparin (60) and the decrease in viscosity and size of hyaluronic acid (66) when incubated with ascorbate at 22°C. The mechanism of this cleavage is not well defined (for review, see 67) but appears to involve free radical formation with oxidative cleavage of glycosidic bonds in the glycosaminoglycan chains. Support for the conclusion that such free radical cleavages of glycosaminoglycans occur in vivo is derived from the demonstration that superoxide systems generate free radicals which depolymerize hyaluronic acid in solution and in synovial fluid at concentrations only 1/3 of that produced by polymorphonuclear leukocytes activated in vitro (68) during phagocytosis.

To see if such molecular cleavage of granule-bound heparin would solubilize heparin from the granule, isolated granules were incubated in the presence and absence of ascorbate and CuSO$_4$. Rat mast cells were gradient purified, incubated for 4 hours in ^{35}S, washed and freeze-thawed six times; and the granules were isolated by differential centrifugation as previously described for immunologically released granules (43). Portions of radiolabeled granules were added to unlabeled granules isolated in the same way. Ten μl of granules containing 25,000 CPM of ^{35}S (approximately 0.75 μg of heparin as determined by metachromasia in the 3 M eluate of Dowex 1 which contained 3/4 of the 68% of the ^{35}S label recovered) and 10 μl of unlabeled granules (approximately 7.4 μg heparin as determined above) were added to 1.5 ml microcentrifuge tubes (Brinkman) containing 0.0075 M Tris, pH 7.4 buffer, alone and with 0.15 M NaCl, 1 M NaCl, and 2.4 mM ascorbate with 4.4 μM CuSO$_4$, respectively, to a final volume of 300 μl. Each mixture was prepared in quadruplicate for determination of ^{35}S-heparin release after 0.5, 1, 2, and 4 hours of incubation at 37°C except in the case of the 1 M

NaCl incubation in which the 2 and 4 hour samples were delet-
ed. An additional sample prepared in Tris buffer alone
served as a zero time control. At the appropriate reaction
time a sample was withdrawn from the incubation mixture,
and the granules were sedimented by centrifugation at 3000 X
g for 20 minutes at 4°C (43). The ^{35}S content of each sample
before centrifugation and in the supernatant after sedimentation of the granules was determined in a scintillation spec-
trometer, and the % ^{35}S-heparin released from the granule was
calculated by the formula:

$$\%^{35}\text{S-heparin released} = \frac{\text{CPM Supernatant}}{\text{CPM Total Sample}} \times 100$$

Net % ^{35}S-heparin released at each time point was equal to %
^{35}S-heparin released, since the zero time control showed no
release of ^{35}S-heparin.

Granules exposed to Tris buffer alone or to buffer with
0.15 M NaCl released 3-16% ^{35}S-heparin over the course of the
incubations (Fig. 4a),

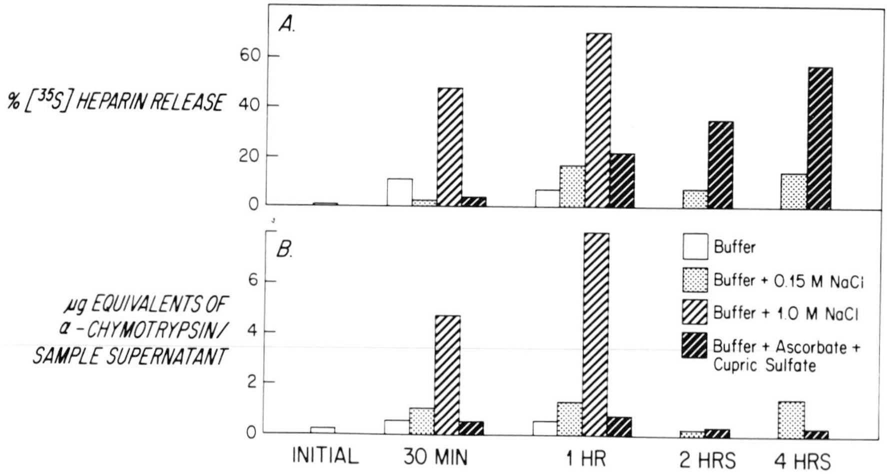

Fig. 4. The release of ^{35}S-heparin (A) and proteolytic
activity (B) from the isolated rat mast cell granule.
Granules were incubated for various time periods at 37°C in
the presence of 0.0075 M Tris, pH 7.4, with 0.15 M NaCl or 1
M NaCl or 2.4 mM ascorbate and 4.4 µM cupric sulfate. The %
^{35}S-heparin released was calculated and proteolytic activity
measured in the supernatant after the granules were sedimented.

while granules exposed to 1 M NaCl for up to 1 hour showed a
time-dependent release of ^{35}S-heparin which filtered at a
molecular weight of approximately 750,000. Similar time-
related results have previously been reported for granules
solubilized in high salt (50) but the size of the released
material was not determined. Granules treated with ascorbate
and CuSO$_4$ showed a slow time-dependent release of ^{35}S-heparin
which reached 56% at 4 hours. The ^{35}S-heparin released at 4
hours by ascorbate and CuSO4 filtered on Sepharose 4B with a
molecular weight of 20,000.

Commercial heparin has a cofactor capacity to augment
markedly the inhibitory activity of antithrombin III on factor
XII low molecular weight fragments (69), thrombin (70), plasma
thromboplastin antecedent (PTA) (71), factor Xa (72), and
plasmin (73) and of the inhibitor of the first component of
complement (ClINH) on the activated first component (74).
Commercial heparin increases diamine oxidase activity in the
plasma of humans (75), and the metachromatic material released
into guinea pig plasma during anaphylaxis increases functional
plasma levels of this histaminase (76). Comparable functional
studies with the high molecular weight native heparin are not
yet available.

B. Proteases

Histochemical studies have demonstrated a chymotrypsin-
like enzyme in the mast cells of the mouse, rabbit, dog,
human (77) and rat (77,78) and a trypsin-like enzyme in mast
cells of dog tissue and human skin (79). Chymotrypsin-like
and trypsin-like enzymes have been extracted from dog mastocytoma (80) and mouse mastocytoma (81). A rat chymotrypsin-
like enzyme termed chymase has been isolated from purified
peritoneal mast cells (82) and localized to the mast cell
granules (40,42,50). The rat mast cell chymase is fully
active on the ester substrate acetyl-tyrosine-ethyl ester in
granule-bound form (44) and the solubilized enzyme is proteolytic for casein (44,82) and albumin (44) substrates. The
chymase has an apparent molecular weight of approximately
25,000 by gel filtration (81) and 29,000 by sodium dodecyl
sulfate electrophoresis (42). The chymase obtained from the
mixed rat peritoneal cell population is inhibited by DFP and
tosylamide phenylmethyl chloromethyl ketone (TPCK) and has
an apparent Km of 1.1×10^{-3} for the specific α-chymotrypsin
substrate, benzoyl tyrosine ethyl ester (BTEE) (82). A similar chymotrypsin-like enzyme by molecular weight and function
has been isolated from mouse mastocytoma (81). A second enzyme
that cleaves p-tosyl-L-arginine methyl ester (TAMe) has also
been isolated from the mouse mastocytoma (81). Mixed human

leukocytes release a TAMe esterase by IgE-dependent mechanisms (83); guinea pig basophil granules, intact and frag-
mented, express distinct TAMe and BTEE esterase (84)
activities which are inhibited by tosyl lysine chloromethyl
ketone and TPCK, respectively; and rat basophils contain a
trypsin-like enzyme as shown by histochemical methods (85).

　　To investigate further the function of this granule-
bound protein component, its esteratic and proteolytic
activities were studied with and without solubilization.
Isolated rat mast cell granules were added in 0.1 ml portions
in 0.0075 M Tris, pH 7.4, to 0.2 ml of the same buffer or to
buffer containing 1.0 M NaCl. The granule samples were incubated for 30 minutes at 25°C and portions were assayed for
their ability to cleave the ester substrate BTEE (86) or for
proteolytic activity on casein. Casein (Hammerstein quality),
labeled by the insolubilized lactoperoxidase method with [125]I
(87) was used as the proteolytic substrate according to the
method of Kunitz as described by Laskowski (88) except that
all volumes were decreased by a factor of 20. A dose dependent cleavage of ^{125}I-casein was obtained between 1 and 10 ng
of α-chymotrypsin (Sigma) and the activity of unknown samples
was determined from the standard curve.

　　Granules treated with buffer alone had more hydrolytic
activity on the small ester than proteolytic activity on the
protein substrate (Fig. 5).

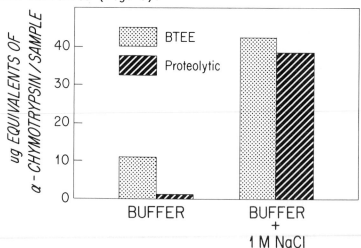

Fig. 5. The esterase and proteolytic activity of isolated rat mast cell granules in a suspension of buffer or
buffer containing 1 M NaCl. Esterase activity was measured
by hydrolysis of the synthetic ester benzoyl-tyrosine-ethyl
ester (BTEE) and proteolytic activity was determined by
cleavage of ^{125}I-labeled casein.

However, granules solubilized with 1 M NaCl exhibited a 3-fold increase in esterase activity, while the proteolytic activity increased more than 30-fold. In order to relate these findings to the solubilization of ^{35}S-heparin from the granules, the fractions from the previous heparin release experiment were assayed for proteolytic activity. The amount of proteolytic activity released from the buffer or the buffer plus 0.15 M NaCl over the total 4 hour incubation varied between 0.3 and 1.5 μg chymotrypsin equivalents (Fig. 4b). Release of activity from the granules incubated with 1 M NaCl reached 4.75 μg at 30 minutes and 8.0 μg at 1 hour and was related to the time course of release of ^{35}S-heparin. No activity was noted in the ascorbate CuSO$_4$-treated samples, but subsequent experiments have shown that α-chymotrypsin is inactivated under such conditions, as have been also found for other enzymes (89,90).

The data shown here (Fig. 4b) indicate that granules treated with 1 M NaCl for 30 minutes released 4.7 μg equivalent of α-chymotrypsin and 48% of the ^{35}S-heparin or approximately 3.9 μg of heparin. In the 1 hour incubation with 1 M NaCl 8 and 5.6 μg of chymase and heparin, respectively, were released. The average of these two weight ratios of 1.3:1 falls within the ratio of granular protein:heparin of 2:1 reported previously (40,91) but is higher than the amount of chymase reported as 15-20% (40,42) of granule protein. The rat mast cell content of heparin is approximately 100 μg/10^6 cells (39), giving a maximum calculated chymase content of 130 μg/10^6 cells; the histamine content is 10-20 μg/10^6 cells (92). Based on molecular weight of these granule components of 750,000, 25,000 and 111 for heparin, chymase and histamine, respectively, the molar ratios are 1:39:676-1352. The 750,000 molecular weight heparin has approximately 2.5 sulfates and 1 carboxyl group per repeating disaccharide unit (molecular weight of approximately 600) (93) and therefore a potential of 4375 anionic binding sites per molecule. The chymase, with approximately 200 amino acid residues, has approximately 16 acidic sites per 100 residues (50,94) or 32 sites per molecule. From the calculated molar ratios the chymase in the granule would offer a maximum of approximately 1200 binding sites compared to approximately 4000 for heparin, of which 1200 would be uronic carboxylic residues. It has been suggested that the O-sulfates are largely committed to binding the basic residues of the highly cationic chymase to form the bulk of the granule (94). The calculated pKa of the granule for histamine in the range of 4-5 (91) implies the appreciable involvement of carboxy residues. The information available is compatible with histamine binding to carboxyl groups in granule-associated protein as suggested by Uvnäs (91) or to such groups in the heparin as favored by Lagunoff (94).

Although histones release a chymotrypsin-like enzyme from rat mast cells (95), current evidence that the rat mast cell chymotrypsin-like enzyme is released by immunologic mechanisms is circumstantial and based on the morphologic (14,23) and chemical evidence (43) that the granule is released. As proposed by others (44,82) and shown here, the granule-bound enzyme, although able to hydrolyze small synthetic substrates, does not express its full proteolytic activity until solubilized. The relationship between the chymotrypsin-like esterase involved in mast cell activation (17,18) and the granular enzyme is not known; however, the finding that α-chymotrypsin releases histamine from the mast cell (22) suggests a potentiating if not a primary action in mast cell degranulation. Crude extracts of dog mastocytoma, containing both tryptic and chymotryptic enzymes, have fibrinolytic and plasminogen activating activity (96). The chymotrypsin-like enzyme is held to participate in the edema formation of a mast cell-mediated axon reflex in the rabbit (97) because the phenomena are blocked by chymotrypsin inhibitors.

The similarity between the rat chymase (82) with respect to molecular weight, chemical inhibition and substrate specificity and the α-chymotrypsin-like enzyme in human peripheral polymorphonuclear leukocytes (98,99) is noteworthy since the latter generates chemotactic activity from serum or C3 or C5 (100), exhibits bacteriocidal activity (101), enhances phagocytosis (102) and the platelet release reaction (103), and degrades rabbit articular cartilage (104).

N-acetyl-β-glucosaminidase has been identified histochemically in rat peritoneal mast cells (105) and localized to the isolated mast cell granule (42,106). This 150,000 molecular weight enzyme (42) is released from rat mast cells after challenge with polymyxin B in a linear relationship to histamine release (107) and has an acidic pH optimum for cleavage of the N-acetyl-β-glucosamine paranitrophenyl substrate (106). The role of this enzyme after release is unknown, although it may act to degrade glycosaminoglycans of the connective tissue ground substance (106), cleave bacterial cell wall into soluble fragments (108), or act on cell surface glycoprotein as shown for blood group substances (109).

C. Eosinophil Chemotactic Factor of Anaphylaxis (ECF-A)

ECF-A was discovered in this laboratory as a 500-1000 molecular weight mediator released by antigen challenge of guinea pig lung slices prepared with IgG₁ (110) and human lung slices sensitized with IgE (5). ECF-A of similar size is released from isolated rat peritoneal mast cells by IgE-dependent mechanisms (37) and by reversed anaphylaxis (111), is granule associated, and is fully liberated from the granule by

low ionicity buffers (37). A chemotactic activity of similar low molecular weight is also releasable from dispersed
concentrated human pulmonary mast cells by IgE-dependent
mechanisms (7) and has been extracted from human leukemic
basophils (112) and mast cell-rich tissues such as nasal
polyps (6).
   The ECF-A tetrapeptides have been synthesized based on
the amino acid sequence of Ala-Gly-Ser-Glu and Val-Gly-Ser-
Glu found in material extracted from human lung (113). The
synthetic tetrapeptide is equal to the purified natural
peptide in in vitro potency and has in vivo activity in
attracting eosinophils into guinea pig peritoneum after
intraperitoneal injection of 0.1 to 1.0 nanomole of either
peptide (114). The N-terminal and C-terminal tripeptide
fragments have minimal chemotactic activity but are inhibitors (113). The N-terminal tripeptide is stimulus-specific
and competes reversibly with the tetrapeptide for a cell
recognition site, whereas the COOH-terminal tripeptide gives
irreversibly cell-directed deactivation. It is suggested
that the tetrapeptides are chemotactic by virtue of binding
of the N-terminal to a hydrophobic domain and activation by
the glutamic acid C-terminal of anionic domain in a receptor
site (115). Both purified native ECF-A and synthetic peptides are preferentially active on eosinophils as compared to
neutrophils and have little activity on monocytes with respect to chemotaxis and the induction of subsequent unre-
sponsiveness to chemotactic stimulation, termed deactivation
(113). Indeed, the tetrapeptides are so potent in eosinophil deactivation that this response may eventually be recognized as their major function. As a concentration gradient
builds up, deactivation would retain cells and later exclude
other cells from a site.
   Two oligopeptides, of molecular weight 2500-3500, with
eosinophilotactic and neutrophilotactic activity have been
found in extracts of purified rat mast cells (111,115) and
are released from these cells by anti-F(ab')$_2$ challenge of
the cells.

D. Neutrophil Chemotactic Factor (NCF)

   NCF was first recognized in extracts of human leukemic
basophils as a high molecular weight factor with preferential
capacity to attract neutrophils (112). Material released
immunologically from rat peritoneal mast cells or extracted
from purified rat mast cells or guinea pig or human lung (37,
115,116) has been found to have preferential neutrophilotactic activity of high molecular weight by gel filtration. A
factor with a molecular weight of greater than 750,000 by gel

filtration with neutrophil chemotactic activity has been
released in association with histamine and a low molecular
weight eosinophilotactic factor into the venous effluent from
cold-challenged extremities of patients with cold urticaria
(117). The relationship of these activities awaits further
definition.

III. PHYSICOCHEMICAL AND FUNCTIONAL CHARACTERISTICS OF UN-
      STORED MEDIATORS

A. Slow Reacting Substance of Anaphylaxis (SRS-A)

     SRS-A is generated after immunologic activation of tissues
(118,119) and cells (120) and was originally characterized by
its contraction of the isolated guinea pig ileum in the presence of mepyramine (118). This base stable (2), sulfur
containing (121) acidic molecule has a molecular weight of
approximately 500 and is inactivated by arylsulfatase B of
the human eosinophil (122) and lung (123) and by arylsulfatases A and B purified from rat leukemic basophils (124).
The mast cell has been presumed to be the cell source of
immunologic generation of SRS-A because the release of SRS-A
follows IgG_1-dependent reactions in guinea pig lung slices
(125) and IgE-dependent reactions in human lung (126) and
nasal polyp fragments (6) and in peripheral leukocyte (127)
and human lung cell suspensions (7,120). In guinea pig lung,
SRS-A generation is prolonged in comparison to histamine
release (118), and in human lung SRS-A generation continues
at a time when the tissue content has plateaued and the release of preformed mediators is complete (120), suggesting
its formation by secondary cell types or by interaction of
antigens with IgG_1 and IgE, respectively, at sites not linked
to the release of preformed mediators (116). Further, SRS-A
generation and release relative to histamine content and
release declines with progressive purification of dispersed
human lung mast cells (7). That more than one cell type and
class of immunoglobulin can participate in SRS-A generation
has been documented by the release of SRS-A into the peritoneal cavity of the rat by either IgE-mast cell (128) or
IgG_a-neutrophil-complement (129) dependent mechanisms.
     In addition to the characteristic contraction of the
mepyraminized guinea pig ileum, partially purified SRS-A con-
tracts only certain isolated smooth muscles (130), impairs the
pulmonary mechanics of the intact anesthetized (131) and un-
anesthetized (132,133) guinea pig, and alters cutaneous vascular permeability in the guinea pig (134). SRS-A is
chemically distinct from the preformed inflammatory mediators

by its base stability, characteristic elution from Amberlite and silicic acid gels (135) and inactivation by arylsulfatases (121-124).

SRS-A-like substances have been generated by calcium ionophore activation of human neutrophils (136) and rat mononuclear leukocytes (137). The generation of this incompletely characterized molecule is apparently decreased by nonsteroidal antiinflammatory drugs (138), whereas the generation of the immunologically generated SRS-A is augmented (139).

B. Platelet Activating Factor (PAF)

PAF activity is measured by its capacity to release ^{14}C-serotonin or other amines from platelets (140). In addition, this activity is associated with the induction of platelet rosettes around rabbit basophils (140) and aggregation of platelets in vitro in the presence of fibrinogen (141). PAF activity was originally recognized to be released from rabbit mixed leukocytes incubated with specific antigen (141-143). However, a more general immunologic association has been found, since this activity has also been derived by antigen challenge of passively sensitized rabbit and human lung (144,145); IgE-mediated reversed immunologic reactions of human mixed leukocytes (142) and rabbit and human lung tissue (144,145); and antigen challenge of the rat peritoneal cavity passively prepared with hyperimmune rat antisera containing predominantly IgGa antibody (146). PAF was not found in extracts of human leukemic basophils before stimulation with the calcium ionophore (112) and this activity increases during incubation of mixed rabbit or human leukocytes in alkaline buffer (142), suggesting that PAF is a generated mediator.

Rat PAF has been purified (146) by applying the chromatographic steps utilized in SRS-A purification (135). This activity cochromatographed with SRS-A and filtered in the same region on Sephadex LH-20. However, the PAF activity was not influenced by treatment with arylsulfatase B and was substantially inactivated by phospholipase D from cabbage and eosinophil sources (146). Rat PAF was not inactivated by phospholipases A or C or by proteolytic enzymes. PAF obtained from rabbit or human mixed leukocytes exhibits cationic characteristics on ion exchange chromatography, filters with a molecular weight of 1100 on Sephadex LH-20, and chromatographs on silica gel thin layer plates in a region staining with phospholipid specific indicators (141,142). This PAF is inactivated by phospholipases A and C but not D (147).

C. Lipid Chemotactic Factors of Anaphylaxis

A lipid chemotactic factor for polymorphonuclear leukocytes is released into the peritoneal cavity of rats passively sensitized with IgG_a and challenged with specific antigen
(148). The deproteinized peritoneal fluid of the challenged
rats was chromatographed on Amberlite XAD-8 with the majority
of the activity eluting with ethanol. The chemotactic activity eluted from a column of silicic acid sequentially developed with hexane, dichloromethane, acetone, propanol and
ethanol:ammonia:water in the acetone fraction; random
migration-enhancing activity was found in the hexane eluate,
and the final solvent characteristically contained SRS-A and
PAF. When polymorphonuclear leukocytes were pretreated with
the acetone eluate, they were deactivated to subsequent chemotactic challenge with that fraction or with C5a. Diffusates
from washed human lung fragments challenged with anti-IgE contained a similar polymorphonuclear chemotactic activity when
purified under the same conditions. Data are not currently
available to demonstrate with certainty that this activity
resides in a generated and unstored mediator(s).

IV. THE CASCADE OF MAST CELL FUNCTION AND ITS REGULATION

The events that follow antigen bridging of IgE antibody
bound to mast cell membranes include membrane perturbation,
intracellular activation, generation of unstored mediators,
and the secretion of newly formed and preformed mediators.
The capacity of the mast cell to modify the tissue environment
is dependent upon at least five levels of regulation, including: intensity of the mast cell activating event; the intracellular controls of generation and release of mediators; the
action of primary and secondary mediators on the mast cell;
the expression of mediator activity in terms of receptor binding and responsiveness of target cells to released mediators;
and the rate at which released mediators undergo biodegrad-
tion (Fig. 6).

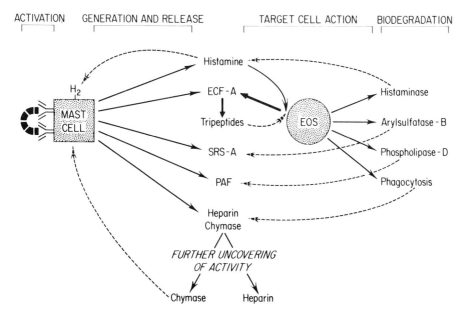

Fig. 6. Schematic representation of five levels of natural regulation of mediator expression.

At the first level, limited IgE-dependent activation of human lung mast cells initiates intracellular accumulation of SRS-A without release to either SRS-A or histamine (120). Further, the capacity of pretreatment with DFP (16) or cytochalasin A or B (119) to suppress IgE-dependent SRS-A generation and release from human lung fragments without impairment of histamine release suggests that SRS-A elaboration involves steps distinct from those of granule-associated release. Colchicine suppresses mediator release without decreasing tissue accumulation of SRS-A (119) and cholinergic agonists have the potential to reverse the inhibitory effect of a beta adrenergic agonist on the release of histamine but not SRS-A. Although control of synthesis of SRS-A may be a major determinant in the accumulation of SRS-A, the presence of arylsulfatase A and B within the cell may contribute to concurrent catabolism of this mediator. Arylsulfatases A and B with full capacity to inactivate SRS-A have been isolated from rat leukemic basophils and identified in purified rat peritoneal mast cells (124).

Elevations in cyclic AMP, achieved synergistically with PGE_2 and aminophylline, synergistically inhibit histamine release (27) from purified rat mast cells. Thus, at the second level both primary mediators and those appearing secondarily could modulate further mediator generation and release. Histamine has been shown to suppress further histamine release in peripheral blood leukocytes enriched for basophils by acting via an H_2 receptor linked to adenyl cyclase (149). The observation that α-chymotrypsin activates the mast cell (22) suggests that the primary release of the chymotrypsin-like enzyme may potentiate mast cell release, particularly in view of the morphologic observation that the granule is frequently found adherent to the plasma membrane (150). Secondary mediators such as PGE_2 and $PGF_{2\alpha}$ can act to suppress and enhance mediator release, respectively, in human lung fragments, presumably due to opposing effects on the cyclic nucleotide system (151). In the cellular phase of the inflammatory response a second activation of the mast cell might occur via the release of cationic protein of the neutrophil (19).

At the third level, some of the mediators, heparin and the chymase in particular, do not fully express activity in their released granule-bound form. Solubilization of the granule and isolation by Dowex chromatography is required to reveal completely the metachromatic and limited anticoagulant activities of native heparin (43,56), since the bound protein partially masks its chemically and biologically active sulfate groups (152,153). Likewise, the granule-bound chymase, although it may have the capacity to act on small peptides such as bradykinin (154), is apparently sterically blocked from protein substrates such as casein. The mechanism of "second activation" of these mediators by solubilization of the granule continues to be unclear, although an intramolecular process such as cleavage of the native heparin in the granule is a distinct possibility.

At the fourth level, inhibitory tripeptides derived from ECF-A by cleavage of the tetrapeptides with carboxypeptidase A and aminopeptidase M (113) which yield COOH-terminal tripeptides and NH_2-terminal tripeptides, respectively, compete with ECF-A for membrane sites (115). The NH_2-terminal tripeptide competes reversibly on an equimolar basis with the stimulus for the eosinophil receptor, whereas COOH-terminal tripeptides irreversibly suppress the cellular eosinophil response. Histamine can either enhance the eosinophil response to ECF-A or at higher concentrations, suppress it, and therefore may also act at this level (115). Another example of target tissue interaction is the ability of SRS-A to potentiate a smooth muscle response to histamine (118).

Both intracellular and extracellular biodegradation, the fifth level, are exemplified by the function of the eosinophil which may facilitate host resistance or prevent adverse effects by limiting the physiologic host response. Although the eosinophil possesses the ability to act in primary host defense via mechanisms such as phagocytosis (155-157) and antibody-dependent cytotoxicity for helminth eggs (158) or larvae (159), a significant regulatory action also appears to be in the degradation, both intracellular and extracellular, of the various mediators. Both heparin and chymase, at least in their granular forms, are phagocytosed by the eosinophil (46-48). The ingested granular matrix apparently is digested by intracellular action (49), but the involvement of eosinophil granular enzymes and the end result of this process are unknown. Eosinophils contain a histaminase capable of oxidative deamination of histamine (160), an arylsulfatase B which can degrade SRS-A (122) and a phospholipase D with the capacity to inactivate rat PAF (146). These enzymes inactivate these mediators in a time- and dose-dependent manner and show coincidence of activity on their respective synthetic and natural substrates through purification, including isoelectric focusing (122,146). SRS-A is inactivated by intact resting eosinophils without release of lysosomal arylsulfatase B (161). Since metabolic inhibitors, which do not affect the isolated enzyme, suppress SRS-A inactivation by the resting eosinophil, SRS-A may be taken up by the cells. Arylsulfatase has been localized to the lamellar bodies of type II (162) pneumocytes of the rabbit lung and found in a broad distribution throughout the cells of dispersed and gradient separated human lung cells (7).

V. REFERENCES

1. Patterson, R., and Suszko, I. M., J. Immunol. 106, 1274 (1971).
2. Orange, R. P., Clin. Allergy 3 (Suppl.), 521 (1973).
3. Selye, H., "The Mast Cell," p. 16, Butterworth, Inc., Washington, D. C., 1965.
4. Feltkamp-Vroon, T. M., Stallman, P. J., Analberse, R. C., and Reerink-Bronkers, E. E., Clin. Immunol. Immunopathol. 4, 392 (1975).
5. Kay, A. B., and Austen, K. F., J. Immunol. 107, 899 (1971).
6. Kaliner, M., Wasserman, S. I., and Austen, K. F., New Engl. J. Med. 289, 277 (1973).
7. Paterson, N. A. M., Wasserman, S. I., Said, J., and Austen, K. F., J. Immunol. 117, 1356 (1976).

8. Hargreave, F. E., and Pepys, J., J. Allergy Clin. Immunol. 50, 157 (1972).
9. McCarthy, D. S., and Pepys, J., Clin. Allergy 1, 415 (1971).
10. Taylor, G., and Shivalkar, P. R., Clin. Allergy 1, 407 (1971).
11. Magro, A. M., and Alexander, A., J. Immunol. 112, 1757 (1974).
12. Siraganian, R. P., Hook, W. A., and Levine, B. B., Immunochemistry 12, 149 (1975).
13. Ishizaka, K., and Ishizaka, T., J. Immunol. 103, 588 (1968).
14. Lawson, D., Fewtrell, C., Gomperts, B., and Raff, M. C., J. Exp. Med. 142, 391 (1975).
15. Becker, K. E., Ishizaka, T., Metzger, H., Ishizaka, K., and Grimeley, P. M., J. Exp. Med. 138, 394 (1973).
16. Kaliner, M., and Austen, K. F., J. Exp. Med. 138, 1077 (1973).
17. Becker, E. L., and Austen, K. F., J. Exp. Med. 124, 379 (1966).
18. Austen, K. F., and Becker, E. L., J. Exp. Med. 124, 397 (1966).
19. Ranadive, N. S., and Cochrane, C. G., J. Immunol. 106, 506 (1971).
20. Uvnäs, B., and Antonsson, J., Biochem. Pharm. 12, 867 (1963).
21. Saeki, K., Jap. J. Pharm. 14, 375 (1964).
22. Lagunoff, D., Chi, E. Y., Wan, H., Biochem. Pharm. 24, 1573 (1975).
23. Anderson, P., Slorach, S. A., and Uvnäs, B., Acta Physiol. Scand. 88, 359 (1973).
24. Perera, B. A. V., and Mongar, J. L., Immunology 6, 472 (1963).
25. Ward, P. A., and Becker, E. L., J. Exp. Med. 125, 1001 (1967).
26. Pearlman, D. S., Ward, P. A., and Becker, E. L., J. Exp. Med. 130, 745 (1969).
27. Kaliner, M., and Austen, K. F., J. Immunol. 112, 664 (1974).
28. Diamant, B., in "Molecular and Biological Aspects of the Acute Allergic Reaction" (S.G.O. Johansson, K. Strandberg, and B. Uvnäs, Eds.), p. 255. Plenum, New York, 1976.
29. Kaliner, M., and Austen, K. F., Biochem. Pharm. 23, 763 (1974).
30. Rölich, P., Exp. Cell Res. 93, 293 (1975).
31. Gillespie, E., and Lichtenstein, L. M., J. Clin. Invest. 51, 294 (1972).

32. Kaliner, M., Orange, R. P., and Austen, K. F., _J. Exp. Med._ 136, 556 (1972).
33. Lawson, D., Raff, M. C., Gomperts, B., Fewtrell, C., and Gilula, N. B., in "Molecular and Biological Aspects of the Acute Allergic Reaction" (S.G.O. Johansson, K. Strandberg, and B. Uvnäs, Eds.), p. 279, Plenum, New York, 1976.
34. Chi, E. Y., Lagunoff, D., and Koehler, J. K., _Proc. Natl. Acad. Sci._ 73, 2823 (1976).
35. Uvnäs, B., and Thon, I. L., in "Mechanisms of Release of Biogenic Amines" (U.S. Von Euler, S. Rosell, and B. Uvnäs, Eds.), p. 361, Pergamon, Oxford, 1966.
36. Lagunoff, D., in "Mechanisms of Release of Biogenic Amines" (U. S. Von Euler, S. Rosell, and B. Uvnäs, Eds.), p. 79, Pergamon, Oxford, 1966.
37. Wasserman, S. I., Goetzl, E. J., and Austen, K. F., _J. Immunol._ 112, 351 (1974).
38. Nosal, R., Slorach, S. A., and Uvnäs, B., _Acta Physiol. Scand._ 80, 215 (1970).
39. Slorach, S. A., _Acta Physiol. Scand._ 82, 91 (1971).
40. Lagunoff, D., Phillips. M. T., Iseri, O. A., and Benditt, E. P., _Lab. Invest._ 13, 1331 (1964).
41. Thon, I. L., and Uvnäs, B., _Acta Physiol. Scand._ 67, 455 (1968).
42. Lagunoff, D., and Pritzl, P., _Arch. Biochem. Biophys._ 1973, (1976).
43. Yurt, R. W., Leid, R. W., Spragg, J., and Austen, K. F., _J. Immunol._ 118, 1201 (1977).
44. Lagunoff, D., and Benditt, E. P., _Ann. N. Y. Acad. Sci._ 103, 185 (1963).
45. Higginbotham, R. D., Dougherty, T. F., and Jee, W. S. S., _Proc. Soc. Exp. Biol. Med._ 92, 256 (1956).
46. Welsh, R. A., and Geer, J. C., _Am. J. Path._ 35, 103 (1959).
47. Mann, P. R., _J. Path._ 98, 183 (1969).
48. Smith, D. E., and Lewis, Y. S., _Anatom. Record_ 132, 93 (1958).
49. Fabian, I., Bleiberg, I., and Aronson, M., _Biochim. et Biophys. Acta_ 437, 122 (1976).
50. Bergqvist, U., Samuelsson, G., and Uvnäs, B., _Acta Physiol. Scand._ 83, 362 (1971).
51. Riley, J. F., and West, G. B., _J. Physiol._ 120, 528 (1953).
52. Schayer, R. W., _Physiol. Rev._ 39, 116 (1959).
53. Black, J. W., Duncan, W. A. M., Durant, C. J., Cannellin, C. R., and Parson, E. M., _Nature_ 236, 385 (1972).
54. Ehrlich, P., _Arch. Anat. Physiol._ 3, 166 (1879).

55. Schiller, S., and Dorfman, A., Biochim. et Biophys. Acta
    31, 278 (1959).
56. Yurt, R. W., Leid, R. W., Austen, K. F., and Silbert,
    J. E., J. Biol. Chem. 252, 518 (1977).
57. Bloom, G., and Ringertz, N. R., Arkiv Kemi 6, 51 (1960).
58. Jansson, L., Ogren, S., and Lindahl, U., Biochem. J. 145,
    53 (1975).
59. Serafini-Francassini, A., Durward, J. J., and Floreani,
    L., Biochem. J. 112, 167 (1969).
60. Horner, A. A., J. Biol. Chem. 246, 231 (1971).
61. Ögren, S., and Lindahl, U., Biochem. J. 125, 1119 (1971).
62. Jaques, L. B., and Charles, A. F., Q. J. Pharmac. 14, 1
    (1941).
63. Engelberg, H., "Heparin Metabolism, Physiology and Clini-
    cal Application," p. 20, Thomas, Springfield, Mass.,
    1963.
64. Orenstein, N. S., Galli, S. J., Hammond, M. E., Smith,
    G. N., Silbert, J. E., and Dvorak, H. F., Fed. Proc.
    36, 5618 (1977).
65. Fillion, G. M. D., Slorach, S. A., and Uvnäs, B., Acta
    Physiol. Scand. 78, 547 (1970).
66. Swann, D., Biochem. J. 114, 819 (1969).
67. Sundblad, L., and Balazs, E. A., in "The Amino Sugars.
    IIB. Metabolism and Interactions" (E. A. Balazs, and R.
    W. Jeanloz, Eds.), p. 229, Academic Press, New York,
    1966.
68. McCord, J. M., Science 185, 529 (1974).
69. Stead, N., Kaplan, A. P., and Rosenberg, R. D., J. Biol.
    Chem. 251, 6481 (1976).
70. Rosenberg, R. D., and Damus, P. S., J. Biol. Chem. 248,
    6490 (1973).
71. Damus, P. S., Hicks, M., and Rosenberg, R. D., Nature
    246, 355 (1973).
72. Yin, E. T., Wessler, S., and Stoll, P. J., J. Biol. Chem.
    246, 3712 (1971).
73. Highsmith, R. F., and Rosenberg, R. D., J. Biol. Chem.
    249, 4335 (1974).
74. Rent, R., Myhrman, R., Fiedel, A., and Gewurz, H., Clin.
    Exp. Immunol. 23, 264 (1976).
75. Tryding, N., Scand. J. Clin. Lab. Invest. 86 (Suppl.),
    196 (1965).
76. Giertz, H., Hahn, F., Krull, P., and Albert, U., Int.
    Arch. Allergy 33, 306 (1968).
77. Gomori, G., J. Histochem. and Cytochem. 1, 469 (1953).
78. Pastan, I., and Almqvist, S., Endocrinology 78, 361
    (1966).
79. Glenner, G. G., and Cohen, L. A., Nature, Lond. 185, 846
    (1960).

80. Katayama, Y., and Ende, N., Nature, Lond. 205, 190 (1965).
81. Vensel, W. H., Komender, J., and Barnard, E. A., Biochim. et Biophys. Acta 235, 395 (1971).
82. Kawiak, J., Vensel, W. H., Komender, J., and Barnard, E. A., Biochim. et Biophys. Acta 235, 172 (1971).
83. Newball, H. H., Talamo, R. C., and Lichtenstein, L. M., Nature 254, 635 (1975).
84. Orenstein, N. S., Hammond, M. E., Dvorak, H. F., and Feder, J., Biochem. Biophys. Res. Comm. 72, 230 (1976).
85. Lagunoff, D., Benditt, E. P., and Watts, R., J. Histo-chem. Cytochem. 10, 672 (1962).
86. "Worthington Enzyme Manual," p. 129, Worthington Bio-chemical Corp. New Jersey, 1972.
87. Thorell, J. I., and Larsson, I., Immunochemistry 11, 203 (1974).
88. Laskowski, M., in "Methods in Enzymology" (S. P. Colowick and N. O. Kaplan, Eds.), Vol. II, p. 33, Academic Press, New York, 1955.
89. Hodgson, E. K., and Fridovich, I., Biochemistry 14, 5294 (1975).
90. Roy, A. B., Biochim. et Biophys. Acta 198, 76 (1970).
91. Uvnäs, B., in "Biochemistry of the Acute Allergic Reac-tions," 2nd International Symp. (K. F. Austen and E. L. Becker, Eds.), p. 177, Blackwell Scientific, Oxford, 1971.
92. Austen, K. F., and Humphrey, J. H., Adv. Immunol. 3, 1 (1963).
93. Lindahl, U., Magnus, H., Bäckström, G., Jacobson, I., Riesenfeld, J., Malmstrom, A., Roden, L., and Feingold, D. S., Fed. Proc. 36, 19 (1977).
94. Lagunoff, D., Biochemistry 13, 3982 (1974).
95. Dekio, S., Ishihara, H., and Yamura, T., Acta Allergolo-gica 31, 193 (1976).
96. Ende, N., and Auditore, J., Am. J. Clin. Path. 36, 16 (1961).
97. Kiernan, J. A., Arch. Derm. Res. 255, 1 (1976).
98. Olsson, I., and Venge, P., Blood 44, 235 (1974).
99. Odeberg, H., Olsson, I., and Venge, P., Lab. Invest. 32, 86 (1975).
100. Venge, R., and Olsson, I., J. Immunol. 115, 1505 (1975).
101. Odeberg, H., and Olsson, I., J. Clin. Invest. 656, 1118 (1975).
102. Hällgren, R., and Venge, P., Inflammation 1, 237 (1976).
103. Hällgren, R., and Venge, P., Inflammation 1, 359 (1976).
104. Malemud, C. J., and Janoff, A., Arth. Rheum. 18, 361 (1975).

105. Pugh, D., and Walker, P. G., J. Histochem. Cytochem. 9, 105 (1961).
106. Lagunoff, D., Pritzl, P., and Mueller, L., Exp. Cell Res. 61, 129 (1970).
107. Lagunoff, D., Biochem. Pharm. 21, 1889 (1972).
108. Schütte, E., and Krisch, K. Z., Physiol. Chem. 311, 121 (1958).
109. Marcus, D. M., Kabat, E. A., and Rosenfield, R. E., J. Exp. Med. 118, 175 (1963).
110. Kay, A. B., Stechschulte, D. J., and Austen, K. F., J. Exp. Med. 133, 602 (1971).
111. Boswell, R. N., Austen, K. F., and Goetzl, E. J., Fed. Proc. 36, 5616 (Abstract) (1977).
112. Lewis, R. A., Goetzl, E. J., Wasserman, S. I., Valone, F. H., Rubin, R. H., and Austen, K. F., J. Immunol. 114, 87 (1975).
113. Goetzl, E. J., and Austen, K. F., Proc. Natl. Acad. Sci. 72, 4123 (1975).
114. Wasserman, S. I., Boswell, R. N., Drazen, J. M., Goetzl, E. J., and Austen, K. F., J. Allergy Clin. Immunol. 57, 190 (Abstract) (1976).
115. Goetzl, E. J., and Austen, K. F., in "Molecular and Biological Aspects of the Acute Allergic Reaction" (S.G.O. Johansson, K. Strandberg, and B. Uvnäs, Eds.), p. 417, Plenum, New York, 1976.
116. Austen, K. F., Wasserman, S. I., and Goetzl, E. J., in "Molecular and Biological Aspects of the Acute Allergic Reaction" (S.G.O. Johansson, K. Strandberg, and B. Uvnäs, Eds.) p. 293, Plenum, New York, 1976.
117. Wasserman, S. I., Soter, N. A., Center, D. M., and Austen, K. F., J. Clin. Invest. 60, 189 (1977).
118. Brocklehurst, W. E., J. Physiol. 151, 416 (1960).
119. Orange, R. P., In "Progress in Immunology II", Vol. 4, p. 29 (L. Brent and J. Holborow, Eds.) North Holland Publishing Co., Amsterdam, 1974.
120. Lewis, R. A., Wasserman, S. I., Goetzl, E. J., and Austen, K. F., J. Exp. Med. 140, 1133 (1974).
121. Orange, R. P., Murphy, R. C., and Austen, K. F., J. Immunol. 113, 316 (1974).
122. Wasserman, S. I., Goetzl, E. J., and Austen, K. F., J. Immunol. 114, 645 (1975).
123. Wasserman, S. I., and Austen, K. F., J. Clin. Invest. 57, 738 (1976).
124. Wasserman, S. I., and Austen, K. F., Fed. Proc. 35, 5612 (Abstract) (1977).
125. Baker, A. R., Bloch, K. J., and Austen, K. F., J. Immunol. 93, 525 (1964).
126. Orange, R. P., Austen, W. G., and Austen, K. F., J. Exp. Med. 134, 136s (1971).

127. Grant, J. A., and Lichtenstein, L. M., J. Immunol. 112, 897 (1974).
128. Orange, R. P., Stechschulte, D. J., and Austen, K. F., J. Immunol. 105, 1087 (1970).
129. Morse, H. C., III, Bloch, K. J., and Austen, K. F., J. Immunol. 101, 658 (1968).
130. Brocklehurst, W. E., Progr. Allerg. 6, 539 (1962).
131. Standberg, K., and Hedqvist, P., Acta Physiol. Scand. 94, 105 (1975).
132. Drazen, J. M., and Austen, K. F., J. Clin. Invest. 53, 1679 (1974).
133. Drazen, J. M., and Austen, K. F., J. Appl. Physiol. 38, 834 (1975).
134. Orange, R. P., Stechschulte, D. J., and Austen, K. F., Fed. Proc. 28, 1710 (1969).
135. Orange, R. P., Murphy, R. C., Karnovsky, M. L., and Austen, K. F., J. Immunol. 110, 760 (1973).
136. Conroy, M. C., Orange, R. P., and Lichtenstein, L. M., J. Immunol. 116, 1677 (1976).
137. Bach, M. K., and Brasher, J. R., J. Immunol. 113, 2040 (1974).
138. Bach, M. K., Brasher, J. R., and Gorman, R. R., Fed. Proc. 36, 565 (Abstract) (1977).
139. Walker, J. L., In "Advances in Biosciences" Vol. 9, p. 235, Pergamon, Vieweg, 1973.
140. Benveniste, J., Henson, P. M., and Cochrane, C. G., J. Exp. Med. 136, 1356 (1972).
141. Benveniste, J., Kamoun, P., and Polonsky, J., Fed. Proc. 35, 985 (Abstract) (1975).
142. Benveniste, J., Nature 249, 581 (1974).
143. Siraganian, R. P., and Osler, A. G., J. Immunol. 106, 1252 (1971).
144. Kravis, T. C., and Henson, P. M., J. Immunol. 115, 1677 (1975).
145. Bogart, D. B., and Stechschulte, D. J., Clin. Res. 22, 652 (Abstract) (1974).
146. Kater, L. A., Goetzl, E. J., and Austen, K. F., J. Clin. Invest. 57, 1173 (1976).
147. Benveniste, J., and Polonsky, J., Fed. Proc. 35, 516 (Abstract) (1976).
148. Valone, F. H., Leid, R. W., and Goetzl, E. J., Fed. Proc. 35, 5615 (Abstract) (1977).
149. Lichtenstein, L. M., and Gillespie, E., J. Pharmacol. Exp. Ther. 192, 441 (1975).
150. Thon, I. L., and Uvnäs, B., Acta Physiol. Scand. 67, 455 (1966).
151. Tauber, A. I., Kaliner, M. A., Stechschulte, D. J., and Austen, K. F., J. Immunol. 111, 27 (1973).

152. Jaques, L., Biochem. J. 37, 189 (1943).
153. Stivala, S. S. and Liberti, P. A., Arch. Biochem. Biophys. 122, 40 (1967).
154. Lagunoff, D., Biochem. Pharmac. Suppl. 17, 221 (1968).
155. Zucker-Franklin, D., Davidson, M., Thomas, L., J. Exp. Med. 124, 521 (1966).
156. Litt, M., J. Cell Biol. 23, 355 (1964).
157. Goetzl, E. J., Wasserman, S. I., Gigli, I., and Austen, K. F. in "New Directions in Asthma" (M. Stein, Ed.) American College of Chest Physicians, Illinois, p. 173, 1975.
158. James, S. L., and Colly, D. G., J. Reticuloendo. Soc. 20, 359 (1976).
159. Butterworth, A. E., Sturrock, R. F., Houba, V., Mahmoud, A. A. F., Sher, A., and Rees, P. H., Nature 256, 727 (1975).
160. Zeiger, R. S., Yurdin, D. L., and Colten, H. R., J. Allergy Clin. Immunol. 58, 172 (1976).
161. Wasserman, S. I., Goetzl, E. J., and Austen, K. F., J. Allergy Clin. Immunol. 55, 72 (Abstract) (1975).
162. DiAugustine, R. P., J. Biol. Chem. 249, 584 (1974).

ACTIVATION OF PLASMINOGEN: A GENERAL MECHANISM FOR PRODUCING LOCALIZED EXTRACELLULAR PROTEOLYSIS

E. Reich

The Rockefeller University, New York, New York   10021

Plasminogen is a protein of molecular weight about 90,000 and it accounts for nearly 0.5% of the plasma in the blood of adult birds and mammals. Plasminogen is itself a proenzyme; its active form, plasmin, is a vigorous general protease of trypsin-like specificity that attacks most proteins, a possible exception being native collagen. The high circulating levels of plasminogen, therefore, represent an enormous reservoir of potential proteolytic activity; this potential could be recruited, either by cells or by self-regulating extracellular processes, for any function requiring localized extracellular proteolysis. Plasminogen was first discovered (1) during a pursuit of the mechanism by which the bacterial activator, streptokinase, stimulates fibrinolysis by plasma and serum; and its suspected role in thrombolysis and in maintaining the fluidity of the blood has been the main focus for interest in the molecule. Although it is reasonable to suppose that plasminogen does indeed participate in the physiology of fibrinolysis, there is as yet no direct evidence either for or against this assumption. On the other hand, the studies briefly summarized in this paper suggest an additional set of functions for the activation of plasminogen; namely, that activation of plasminogen provides a general mechanism, perhaps the general mechanism, for

generating proteolysis localized to the immediately surrounding cellular microenvironment under normal and
pathological conditions.

The discovery that cells can produce fibrinolytic
activity was made by Carrel and Burrows in 1911 (2).
They observed that explanted fragments of a sarcoma
dissolved the plasma clots which provided the semi-solid
support for tissue culture at the time. This observation was later repeated and extensively documented by
Fischer (3); he correlated rapid clot lysis with neoplastic states of explanted tissues, but did not succeed
in demonstrating either the cellular or the enzymatic
basis of the phenomenon and his contributions were ul-
timately consigned to undeserved neglect. Since the
clot lysis is an enzymatic process, Fischer's work suggested the existence of an enzymatic difference between
normal tissue and its malignant derivatives, and the
important implications of this possibility spurred us to
re-examine his results and conclusions.

Tumor Associated Fibrinolysis: Plasminogen Activator Secretion by Virus Transformed and by Other Malignant Cells

Chick embryo fibroblasts were selected as the first
material to test whether Fischer's observations could be
repeated under the conditions of modern, quantitative
cell culture, since Fischer initially compared the fibrinolytic activity of normal chick connective tissue
with that of sarcomas induced by the Rous sarcoma virus
(RSV). By culturing cells on fibrin films both visual
and quantitative confirmation of Fischer's observations
was obtained (4): normal fibroblast cultures produced
little fibrinolytic activity, the fibrinolytic activity
of RSV-transformed cultures was at least 20-fold higher.
A survey of different lytic, temperate, and oncogenic
viruses showed that enhanced fibrinolysis was observed
only when cells were infected with transforming sarcoma
viruses that were also tumorigenic in chicks; mere cell
lysis or infection with lytic DNA or RNA viruses did not
lead to increased fibrinolysis.

Several findings permitted a more detailed description of the relationship between fibrinolysis and viral transformation. The first of these was the development of a cell free assay system following the demonstration that fibrinolysis in cell culture required the interaction of two factors, one being contributed by the serum supplement in the growth medium, and the other by the cells (4). Since the cell factor was secreted and accumulated in the medium, fibrinolysis could be produced simply by mixing serum-free conditioned medium with serum, and measuring the resulting enzyme activity by its solubilizing action on ^{125}I-fibrin. With such a cell-free assay available both the cell and serum factors could be isolated and characterized using routine procedures for protein purification; in this way the serum factor was identified as the proenzyme plasminogen (5), and the cell factor as a limited arginine-specific serine protease that activated plasminogen to plasmin (6). The differences in fibrinolytic activity between transformed and control cultures simply reflected the differences in production of the cell factor, plasminogen activator (6).

A second finding of importance was that, apart from appearing in the cell culture fluids, plasminogen activator also accumulated within RSV-transformed cells, but not in normal uninfected controls (6). Persuasive evidence of the close association between viral transformation and plasminogen activator production was obtained by the use of RSV mutants that are temperature sensitive for transformation: (a) Both extracellular (4) and intracellular (7) levels of enzyme changed and reflected the expression of the transforming viral genes. These changes required both mRNA and protein synthesis, indicating that induction and deinduction were occurring at the level of genetic regulation. (b) The magnitude of these effects on enzyme production (10-100 fold) (4,7,8) make plasminogen activator unique among the biochemical changes that accompany oncogenic transformation. (c) Changes in plasminogen activator formation occurred very rapidly: the intracellular pool reflected these within 2 hr, and secreted enzyme shortly thereafter. (d) The catalytic specificity of the plasminogen activator produced in virus transformed cells was determined by the host cell and not by the virus (9).

These results showed that alterations in plasminogen activator production are cellular responses to the
changing expression of transforming viral genes; these
responses are initiated at the genetic level and they
require both genetic transcription and translation, but
there is still no clear proof that the changes in regulation directly involve the structural gene for plasminogen activator itself.

How general is the association between neoplasia
and plasminogen activator? Without any exception, primary and early subcultures of chick, mouse, rat, hamster
and human embryonic cells produce very low levels of
this enzyme, and transformation by oncogenic DNA or RNA
viruses or by carcinogenic hydrocarbons is accompanied
by large increases in enzyme synthesis (4,7,8,9,10,11,
12). Likewise, transformation of differentiating myo-
genic cultures after infection by RSV is associated with
increases in plasminogen activator synthesis comparable
to those found in transformed fibroblast cultures (13).

Although the available data are still limited,
equally significant correlations have emerged from analysis of enzyme content of primary tumors in animals.
As originally reported by Fischer, Rous sarcomas obtained from virus-infected birds always secrete very
high levels of fibrinolytic activity (10) as do primary
hepatomas of animals fed the chemical carcinogen acetyl-
aminofluorene (11) and the skin tumors of mice initiated
with 7,12-dimethylbenzanthracene (11). In a series of
over one hundred spontaneous rat mammary tumors the
only ones showing high rates of proteolysis were those
diagnosed as malignant by morphological criteria, while
approximately one hundred malignant mammary tumors of
viral (mammary tumor virus in mice) or chemical origin
(nitrosomethylurea or dimethylbenzanthracene in rats)
invariably contained very high concentrations of plasminogen activator (11). These observations are rein-
forced by the finding that when loss of tumorigenicity
in melanoma cell lines is induced by exposure to 5-
bromo-2'-deoxyuridine, there is an accompanying disappearance of high rates of plasminogen activator production (14). Hence, the correlation between elevated
plasminogen activator levels and malignant transformation appears so far to be excellent.

The preceding correlations have been strongly re-
inforced by work on tumor promoting substances such as
phorbol esters and retinoids. Troll and his colleagues
have correlated proteolysis with tumorigenesis in mouse
skin by showing (a) that increased protease activity
appeared rapidly after topical application of phorbol
ester, and (b) that glucocorticoid blocked both the
tumor-promoting and protease-inducing action of phorbol
esters (15). More recently, nanomolar concentrations
of phorbol esters were shown to induce high levels of
plasminogen activator production by cultured chick embryo fibroblasts and HeLa cells (16), and by macrophages
(17), granulocytes (18), and epidermal cells (19); in
the latter three cases the effect of phorbol ester in
cell culture was blocked by glucocorticoids, just as it
is on mouse skin in vivo. A further correlation between
tumor promotion and plasminogen activator synthesis was
found with retinoids, which stimulate the growth of viral sarcomas in fowl (20); physiological concentrations
also induced high rates of enzyme production by cultured
chick cells (21). A striking observation in these studies (16,21) was that cells became both hypersensitive
and hyperresponsive to the inducing actions of phorbol
esters and retinoids (21) after transformation by oncogenic viruses, their behavior in this respect mimick-
ing the known hyperresponsiveness of neoplastic tissue
to hormones. Since the action of promoting agents is
defined by their effects at the onset of tumor development these findings imply that enhanced plasminogen
activator production is associated with malignancy from
the earliest stages of tumor growth. During the initial
phase enzyme synthesis would be dependent on promoting
stimuli, whereas the autonomy of the fully developed
tumor is reflected in the independent or constitutive
expression of this function.
        What is the nature of the oncogenic stimulus to
which the cell responds by producing plasminogen activator? And what is the significance of enzyme production for the phenotype of malignant cells? While we
are only beginning to explore for answers to the first
of these questions, the second has governed most of our
work during the past few years. Two of the characteristics of malignant tumors appear potentially related

to enhanced plasminogen activator production--firstly, the ability of tumors to invade locally, to migrate, and to implant at distant sites, and secondly, the abnormal tissue architecture which provides the morphological basis for tumor identification and pathological diagnosis. Neither of these processes is limited to tumors since many normal cells are widely invasive during embryonic and adult life. We have, therefore, surveyed a number of cellular events that can be regarded as normal models of the malignant phenotype to search for possible biochemical unity in the enzymatic correlates of tissue remodeling and cell migration: the evidence obtained to date strongly supports the assumption that such unity does exist.

Plasminogen Activator Production by Inflammatory Cells

In their ability to cross tissue barriers and to migrate throughout the body, macrophages and granulocytes strongly resemble invasive malignant cells. To enter the bloodstream after maturation in hemopoietic tissues, inflammatory cells must create discontinuities that allow them to emerge from the marrow by penetrating between adjacent endothelial cells; the same process is repeated when these cells escape from the circulation as they are recruited to sites of inflammation in the tissues. Hence, in the course of their normal life cycle, both macrophages and granulocytes reproduce the cellular behavior associated with the hematogenous dissemination of tumor metastases. Moreover, these cells are also attractive as experimental models for defining biochemical correlates of migration since glucocorticoids and a variety of inflammatory agents are known to modulate their migration both in vivo and in vitro.
Plasminogen activator production by mouse macrophages and human granulocytes is closely correlated with the response of these cells to modulators of inflammation (17,18,22,23):
(a) Peritoneal macrophages obtained from mice that have been injected with inflammatory substances such as thioglycollate medium, endotoxin, mineral oil, etc., produce large amounts of plasminogen activator, whereas

the resident peritoneal macrophages from unstimulated
animals do not (24,25). Enzyme synthesis is also induced in a dose-dependent manner when macrophages are
exposed to asbestos either in vivo or in vitro (22).
Under these conditions there are no concurrent changes
in other macrophage enzymes such as lysozyme or lysosomal hydrolases.

(b) The production of plasminogen activator by
macrophages and granulocytes can be induced in vitro by
the potent inflammatory tumor promoter phorbol myristate
acetate and by specific lectins such as concanavalin A.

(c) Plasminogen activator production by inflammatory cells can be inhibited by anti-inflammatory agents
such as glucocorticoids and anti-mitotic compounds (23).
The inhibitory effect of glucocorticoids is profound,
rapid in onset, reversible, and it occurs at hormone
concentrations comparable to those present in the body
fluids under physiological conditions. The order of
relative inhibitory potencies of glucocorticoids exactly
reflects their respective anti-inflammatory potencies in
vivo, and their action on plasminogen activator production is rather selective since this was the only function to be affected among a large number tested.

(d) Glucocorticoid administration rapidly produces
a decrease in circulating monocytes and prevents the
appearance of macrophages (and granulocytes) in inflammatory exudates (26). Since glucocorticoids do not
block the multiplication of leukocyte precursors, these
facts are an indication that glucocorticoids prevent
the migration of inflammatory cells from the marrow into
the circulation, and then from the circulation to sites
of inflammation.

Taken together, all of the preceding correlations
suggest the following working hypothesis: that in association with their migration between the various body
compartments, cells require an enzymatic mechanism that
digests some of the supporting structure of blood vessels and connective tissue. It is proposed that plasmin, by acting on the adhesive substances on the surface
of endothelial cells, on the connective tissue structure
of small blood vessels, and on the other components of
the extracellular matrix, provides an essential part of
the enzymatic basis for migration, and that plasmin is

generated locally by cellular synthesis and secretion of
plasminogen activator. In this view of migration,
translocation of cells would be dynamically integrated
with concurrent microscopic tissue remodeling. By
blocking enzyme production, the administration of glucocorticoids would, to a large extent, prevent migration
of inflammatory cells, and hence suppress the cellular
component of inflammation. Even if this hypothesis
should prove to be correct, it should not be taken to
imply that the anti-inflammatory effects of steroids are
entirely attributable to repression of plasminogen activator synthesis.

Plasminogen Activator Production by Trophoblast

The early development of vertebrate embryos offers
numerous examples of tissue remodeling and cell migration, and thus provides experimental opportunities for
testing the extent to which these processes are correlated with cellular production of plasminogen activator.
The trophoblast is the cell type of greatest interest
for this purpose since it is well established that trophoblast cells invade the uterine wall or ectopic sites
during embryo implantation (27), and they provide one
classic example of normal cells that are responsible for
gross tissue disruption. A study of mouse embryos during early stages of development gave the following results (28): (a) Plasminogen activator production was
undetectable prior to differentiation of the morula into
blastocyst. (b) All blastocysts produced plasminogen
activator, and the enzyme was present intracellularly as
well as in conditioned medium. (c) The onset of enzyme
production was a function of equivalent gestation day
(i.e., age of embryo) and not of time in culture. (d)
The trophoblast was responsible for all of the enzyme
initially produced in culture, the time course of enzyme production by pure cultures of trophoblast cells
coincided with the invasive phase of their cellular
lifespan in utero (Fig. 1).
These results showed that plasminogen activator was
produced by trophoblast and that its production was precisely regulated. The trophoblast is one of the most

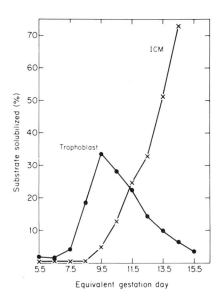

Fig. 1. Production of plasminogen activator by cultures of isolated trophoblast and inner cell masses. Each curve represents the average of two separate experiments. For experimental details see ref. 28. Reprinted by permission from Cell.

invasive cells known because it can invade and destroy virtually all tissues in addition to those normally encountered in the uterus, and it has frequently been compared with malignant tumor cells, but its invasiveness is expressed only during a limited period of its life cycle. The invasive period has been determined primarily from transplantation of early embryos to ectopic sites, and it extends from the sixth to the tenth equivalent day of gestation in the mouse (27). This interval corresponds closely with the period during which cultured trophoblast is secreting plasminogen activator. In fact, since the invasiveness of the trophoblast is arrested by mid-gestation, it is of interest that trophoblast dissected from tenth day conceptuses no longer

produce significant amounts of enzyme. All of these
findings suggest that plasminogen activator secretion
is a contributing factor in implantation.

Ovarian Plasminogen Activator: Relationship to Ovulation and Hormonal Regulation

The mature ovarian follicle in mammals consists of
a fluid-filled cavity surrounded by granulosa cells to
which the ovum and its associated structures are attached. The granulosa cells are bounded by a basement
membrane which, in turn, is encapsulated by a thick and
dense connective tissue--the theca (Fig. 2). In order

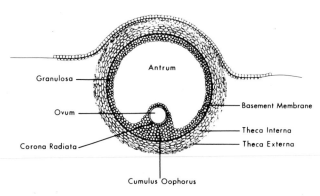

Fig. 2. Structure of mammalian ovarian follicle.

for the ovum to escape from this structure and enter the
oviduct where it is fertilized, extensive degradation of
the follicle wall is required, and the events that occur
at ovulation are one of the most dramatic examples of
tissue remodeling in adult mammals. Many years ago
Schochet (29) compared the translocation of the ovum at
ovulation to tumor cell metastsis and proposed that proteolysis within the follicle was a likely mechanism for
disrupting the follicle wall. More recent work had also

shown that the morphological changes that follow intra-
follicular injection of trypsin resemble those that occur at ovulation (30). However, it remained to be demonstrated that generation of proteolytic activity was
actually correlated with ovulation.

The ubiquity of plasminogen and plasminogen activators, and the presence of plasminogen in follicular
fluid (31) suggested a possible role for these enzymes
in ovulation. An experimental system suitable for testing this hypothesis had been developed by Crisp and
Denys (32) who showed that pure cultures of granulosa
cells could be obtained by puncturing follicles with
fine needles; hormonally synchronized induction of ovulation in immature animals allowed sufficient material
to be prepared for biochemical analysis from precisely
timed stages of follicular development. The following
summary outlines the relationship between ovulation and
plasminogen activator production by granulosa cells (33,
34): (a) Granulosa cells isolated from follicles eight
hours or longer prior to the expected time of ovulation
produce little or no detectable plasminogen activator.
(b) Both the numbers of granulosa cells producing high
levels of plasminogen activator, and the rate of enzyme
production increase rapidly during the 8 hr interval
immediately preceding ovulation. (c) Cultured granulosa
cells obtained from follicles not destined to ovulate do
not spontaneously produce plasminogen activator, but
they may be induced to synthesize and secrete large
amounts of enzyme by exposure either to picomolar concentrations of gonadotrophic hormones, or to cyclic nucleotides, or to some prostaglandins. (d) Granulosa
cell cultures prepared from the ovaries of sexually
mature, normally cycling rats produce significant quantities of plasminogen activator only when harvested from
follicles on the night of proestrus, that is, immediately before the expected time of ovulation. Cells obtained from animals in estrus or diestrus produce little
or no enzyme.

All of the results demonstrate a close correlation
between the level of granulosa cell plasminogen activator and ovulation. The correlation includes hormonal
determination of the two processes, and their temporal
relationship both in normally cycling mature animals,

and in immature animals primed with hormones. These
findings suggest the following hypothesis for the enzymatic basis of the events that culminate in ovulation:
(a) the increased circulating levels of gonadotrophic
hormone that occur before ovulation stimulate granulosa
cells to increase their intracellular concentration of
cAMP; (b) in response to the rise in cAMP granulosa
cells synthesize plasminogen activator and secrete this
enzyme into the follicular fluid; (c) once in the follicular fluid, plasminogen activator catalyzes the formation of plasmin which degrades the follicle wall by
proteolysis, liberating the ovum.

These conclusions, and the experimental evidence
on which they are based, provide gratifying support for
Schochet's suggestions concerning both the enzymatic
requirements for ovulation and the comparison between
ovulation and tumor cell migration.

Conclusions

As outlined briefly in the preceding summary, processes seemingly as diverse as inflammation, ovulation,
trophoblast implantation and neoplasia are related by
structural, cellular, and enzymatic correlates. The
common elements in these phenomena that have been emphasized here are tissue remodeling and cell migration;
and, given the nature of the processes involved, it is
not surprising that elevated rates of localized extracellular proteolysis, initiated by plasminogen activator
secretion, should be prominently associated in each
case. Moreover, the pattern of plasminogen activator
synthesis qualitatively mirrors the patterns of cellular
regulation characteristic of each process: enzyme production is transient and it is precisely regulated,
either by temporal programming or by hormonal controls
during normal tissue function in ovulation, inflammation and implantation, whereas it appears to be continuous and constitutive in malignant tissue. Although
envisioning parallels between inflammation, ovulation
and implantation on the one hand, and neoplasia on the
other, may seem fanciful, the common features in these
tissue reactions have been recognized by pathologists

for many years, and they are not limited to the factors discussed above. Other similarities include changes in the structure and permeability of the local vasculature, and a conspicuous parallelism between the inflammatory and tumor promoting potencies of the diterpene co-carcinogens related to phorbol. It is reasonable to expect that biochemical analysis will yield additional examples of common and correlated enzymatic changes in these and other analogous tissue responses.

Summary

Plasminogen activator is synthesized by most vertebrate tissues, and it appears to be a general mechanism for producing localized extracellular proteolysis. Elevated rates of enzyme secretion are characteristically observed in association with processes that require proteolysis in the local microenvironment; these include tissue remodeling and cell migration in normal tissue functions such as embryonic development, ovulation, blastocyst implantation, and inflammation, in which plasminogen activator synthesis is transient and precisely regulated. Enzyme synthesis in neoplastic cells appears to be constitutive and permanently deregulated, at least in part.

References

1. Christensen, L.R., and MacLeod, C.M., J. Gen. Physiol. 28, 559 (1945).
1a. Astrup, T., in "Proteases and Biological Control" (E. Reich, D.B. Rifkin and E. Shaw, eds.) Cold Spring Harbor Conferences on Cell Proliferation, Cold Spring Harbor, pp. 343-355. 1975.
2. Carrel, A., and Burrows, M.T., J. Exp. Med. 13, 571 (1911).
3. Fischer, A., Arch. Entwicklungs 104, 210 (1925).
4. Unkeless, J.C., Tobia, A., Ossowski, L., Quigley, J.P., Rifkin, D.B., and Reich, E., J. Exp. Med. 137, 85 (1973).
5. Quigley, J.P., Ossowski, L., and Reich, E., J.

Biol. Chem. 249, 4306 (1974).

6. Unkeless, J.C., Danø, K., Kellerman, G., and Reich, E., J. Biol. Chem. 249, 4295 (1974).

7. Rifkin, D.B., Beal, L., and Reich, E., in "Proteases and Biological Control" (E. Reich, D.B. Rifkin and E. Shaw, eds.) Cold Spring Harbor Conferences on Cell Proliferation, Cold Spring Harbor, pp. 841-847. 1975.

8. Danø, K., and Reich, E., J. Exp. Med. 147, 745 (1978).

9. Ossowski, L., Unkeless, J.C., Tobia, A., Quigley, J.P., Rifkin, D.B., and Reich, E., J. Exp. Med. 137, 112 (1973).

10. Ossowski, L., and Reich, E., In preparation.

11. Piperno, A., Tobia, A., Mouradia, J., Rifkin, D.B., and Reich, E., Unpublished observations.

12. Barrett, J.C., Crawford, B.D., Grady, D.L., Hester, L.D., Jones, P.A., Benedict, W.F., and Tso, P.O.P. Cancer Res. 37, 3815 (1977).

13. Miskin, R., Easton, T.G., and Reich, E., Submitted to Cell.

14. Christman, J.K., Acs, G., Silagi, S., and Silverstein, S., in "Proteases and Biological Control" (E. Reich, D.B. Rifkin and E. Shaw, eds.) Cold Spring Harbor Conferences on Cell Proliferation, Cold Spring Harbor, pp. 827-840. 1975.

15. Troll, W., Rossman, T., Katz, J., Levitz, M., and Sugimura, T., in "Proteases and Biological Control" (E. Reich, D.B. Rifkin and E. Shaw, eds.) Cold Spring Harbor Conferences on Cell Proliferation, Cold Spring Harbor, pp. 977-987. 1975.

16. Wigler, M., and Weinstein, I.B., Nature 259, 232 (1976).

17. Vassalli, J.-D., Hamilton, J., and Reich, E., J. Exp. Med. 146, 1693 (1978).

18. Granelli-Piperno, A., Vassalli, J.-D., and Reich, E., J. Exp. Med. 146, 1693 (1978).

19. DiGirolamo, P., Green, H., and Reich, E., unpublished observations.

20. Polliack, A., and Sarson, Z.B. J. Nat. Cancer Inst. 48, 407 (1972).

21. Wilson, E.L., and Reich, E., Submitted to Cell.

22. Hamilton, J., Vassalli, J.-D., and Reich, E., J.

Exp. Med. 144, 1689 (1976).
23. Vassalli, J.-D., Hamilton, J., and Reich, E., Cell 8, 271 (1976).
24. Unkeless, J.C., Gordon, S., and Reich, E., J. Exp. Med. 139, 834 (1974).
25. Gordon, S., Unkeless, J.C., and Cohn, Z., J. Exp. Med. 140, 995 (1974).
26. Thompson, J., and Van Furth, R., in "Mononuclear Phagocytes" (Ralph Van Furth, ed.) Blackwell Scientific Publications, Ltd. Oxford. pp. 255-264. 1970.
27. Kirby, D.R.S., in "The Early Conceptus, Normal and Abnormal" (W.W. Park, ed.) University of St. Andrews Press. Edinburgh. pp. 68-73. 1965.
28. Strickland, S., Reich, E., and Sherman, M.I., Cell 9, 213 (1976).
29. Schochet, S.S., Anat. Rec. 10, 447 (1916).
30. Espey, L.L., Biol. Reprod. 10, 216 (1974).
31. Beers, W.H., Cell 6, 381 (1975).
32. Crisp, T.M., and Denys F.R., in "Electron Microscopic Concepts of Secretion" (M. Hess, ed.) John Wiley and Sons. New York. pp. 3-33. 1975.
33. Beers, W., Strickland, and Reich, E., Cell 6, 387 (1975).
34. Strickland, S., and Beers, W., J. Biol. Chem. 251, 5694 (1976).

PROTEOLYTIC EVENTS IN VIRAL REPLICATION

Bruce D. Korant

E. I. du Pont de Nemours and Company

The role of protein processing in animal virus and bacteriophage replication is reviewed. Examples are used from various virus groups of protein cleavage in regulation of viral assembly, maturation and genome replication. Evidence is presented for a role of host cell proteases, and for induction of viral-specific proteases. The structures of cleavage sites in viral precursor proteins are compared. Inhibition of viral proteolysis by alteration of the substrates and by inactivation of proteases is described.

The successful replication in susceptible host cells of animal viruses requires the action of proteolytic enzymes on viral precursor proteins. The protein cleavages are often crucial to the virus replication process, so that a complete understanding of their role could lead to development of new types of antiviral agents. On a more fundamental level, the characterization of the proteolytic reactions in virus infections is leading to a fuller appreciation of regulation of function by limited proteolysis. Viruses provide as well, models in which proteolytic modifications yield macromolecular complexes with altered conformational and antigenic properties, and altered affinities for nucleic acids and receptors on cell membranes. Obviously, there is a general significance of these

171

phenomena in understanding similar events in cellular
interactions and in the numerous proteolytic events
occurring in the tissues and fluids of higher organisms.
    There have been several reviews of proteolysis in
virus replication (1,2,3,4), but all are incomplete due
to a large number of significant recent reports, particularly with the tumor viruses. In addition, where
it is relevant to a comparison with animal viruses, I
will include references to studies on protein cleavage
in bacterial virus replication. There are now several
examples of recent progress in characterization of the
specificity of the reactions, the origin of the proteases and the role of the cleavage as a control element. I plan to use this review to discuss these
points in as much detail as current information permits.

I. EXTENT OF PROTEIN CLEAVAGE AMONG VIRUSES

    The animal virus groups in which proteolytic clea-
vage has been identified as a process in replication
are listed in Table 1. It is obvious that many diverse
types of viruses, with little or no genetic relationship to one another, display the phenomena of proteolytic processing. Indicated in Table 1 is the fact
that many of these viruses have proteins cleaved during
assembly of subunits into larger aggregates, or in the
maturation process when the viral genome is finally
combined with capsid proteins to yield an infectious
virion. Although proteolysis accompanying maturation
is the most common theme, there are other variations in
the requirements for proteolytic processing during
early stages of infection, or the replication process
itself.
    The basic cycle of virus infection of susceptible
cells involves recognition of the cell surface by some
determinant on the virus capsid surface, binding, followed by a destabilizing alteration of the infecting
virion leading to the introduction of the viral nucleic
acid into the cell. Once present, the viral genome is
amplified by a set of replicases (polymerases). These
may be present in the cell prior to infection, carried

TABLE I   Characteristics of Virus Group
in Which Protein Cleavage Occurs

Virus	Structural Features of Virion	Role of Cleavage
Picorna Polio	SS(+)[1] RNA, protein coat	All viral proteins are cleavage products, most functions regulated by proteolysis
Toga Sindbis	SS(+) RNA, protein core, glycoprotein-lipid envelope	Similar to picorna
Oncorna (Retro) Rous sarcoma	SS(+) RNS, protein core, reverse transcriptase, glycoprotein-lipid envelope	Probably similar to picorna
Reo	DS RNA, protein core, outer protein coat, RNA transcriptase	Activation of transcriptase, maturation
Myxo Influenza	SS(-) RNA, protein core, glycoproteins (hemagglutinin and neuraminidase) and lipid envelope, RNA transcriptase	Activation of virions (hemagglutinin)
Paramyxo Sendai	Similar to myxo	Absolute requirement that envelope glycoprotein be cleaved for infectivity
Adeno	DS DNA, numerous protein structures in capsid	Maturation
Pox Vaccinia	DS DNA, many proteins and lipid in capsid	Maturation
Herpes simplex	DS DNA, protein core proteins and lipid in envelope	?
Papova SV40	DS DNA, three proteins in capsid, histones in core	Capsid protein synthesis, T antigen production (transformation)
Parvo Adeno-associated virus	SS(+) DNA, three proteins	Capsid protein synthesis and assembly

[1]*Abbreviations: SS-single stranded, DS-double stranded, (+)-infectious (messenger RNA-like), (-)-complementary, transcribed into messenger RNA.*

in by the infecting virus, or produced as translation
products of the viral genome. Viral messenger RNA is
synthesized and utilizes cell ribosomes and parts of
the cellular translation apparatus to produce viral
proteins. Some of these proteins are enzymatic, and
participate further in production of progeny viral
genomes. Other of the proteins are structural elements
which assemble and combine with the viral nucleic acid
to form progeny virus. These exit the cell. Often,
but not always, the infected cell is killed, and lysis
accompanies or follows virion release. In other
examples, the cells survive, and may continue to produce virus. In the extreme case, the cell may be
transformed by the infection, and yield a potentially
"immortal" line which conserves a portion of the viral
genes and displays new characteristics reflective of
the infecting virus. The cells continue to divide and
may express viral antigens or occasionally yield virions. These diverse and intricate features of the cycle
of virus replication contain examples of proteolysis as
a control function. Some of these are indicated
diagrammatically in figure 1.

Discussion of the role of proteolysis in some of
the processes of replication will follow. For details
within a virus group not covered in depth, the reader
is referred to other reviews (1,4,5).

II.   POLYPROTEINS OF PICORNA AND TOGAVIRUSES--PROTEO-
      LYTIC REGULATION OF ALL VIRAL FUNCTIONS

A. Picornavirus Protein Cleavage

The picornaviruses are the prototype for studies
on animal virus protein processing (6,7,8). A large
number of laboratories have confirmed and extended the
initial reports, and a clear, although still incomplete
picture has emerged of these viruses whose proteins are
virtually all produced by proteolysis. A schematic is
shown in figure 2 which summarizes the important stages
in the processing reactions.

The viral messenger RNA codes for approx. 200 Kd
of protein. In vivo there appears to be one initiation
site for translation, and starting at this point translation begins, apparently of a sequence adjacent to the
amino terminus of the coat protein gene (8,9,10). As
the ribosome leaves the region of the structural gene,
a cleavage of the nascent chain occurs. Nevertheless,
the ribosome proceeds, without reinitiating, and trans-
lates the next region, coding for polypeptide X. A
second nascent cleavage occurs, but again the ribosome
proceeds and completes the replicase gene product (9,
11-13). Since these cleavages, which are carried out
by a host cell protease(s) (see below) occur during
translation, actually little or no polyprotein is normally observed. Special inhibitory treatments, such as
amino acid analogs or protease inhibitors, are required
to clearly demonstrate the species larger than the coat
precursor of 95 Kd (see gel pattern C, in figure 2).
The transit time of the ribosome through the structural
gene region is only fractions of a minute. This indicates that there is little time for recognition of a
conformational feature of the precursor. Rather, the
signal must be a simple one built into the primary
sequence of the polyprotein. In essence, the cleavages
of the nascent picornavirus chains resemble those
removing precursor regions of prohormones (14) and
other secretory proteins (15), and the viruses may well
be utilizing parts of the same processing system.

Additional proteolytic reactions produce the stable structural polypeptides. The cleavages take
several minutes to occur (16), and lead to extensive
refolding of the structural precursor as it assembles.
An interesting aspect of these reactions is that the
proteases responsible are not active in uninfected
cells and the enzymes are probably virus-coded. Processing of the RNA replicase polypeptides takes much
longer (15 min. or more at the midcycle of infection),
and provides a control over the action of viral RNA
synthesis (see below). Processing reactions leading to
production of stable picornavirus proteins may be
slowed by treatments which do not delay the nascent
cleavages (fig. 2, gel pattern b). This implies that

PROCESS

1. VIRUS-CELL INTERACTIONS
   a. ATTACHMENT
   b. PENETRATION
   c. RELEASE OF NUCLEIC ACID

2. INHIBITION OF HOST CELL SYNTHESIS

3. VIRAL PROTEIN SYNTHESIS

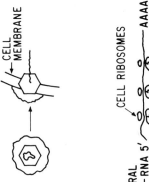

4. REGULATION OF NUCLEIC ACID SYNTHESIS

5. CAPSOMERE ASSEMBLY

EXAMPLES

PARAMYXO, MYXO, REO, PICORNA (?)

PICORNA

PICORNA, TOGA, ONCORNA, ADENO

REO, TOGA, ONCORNA, PICORNA, HERPES (?)

MOST VIRUS GROUPS

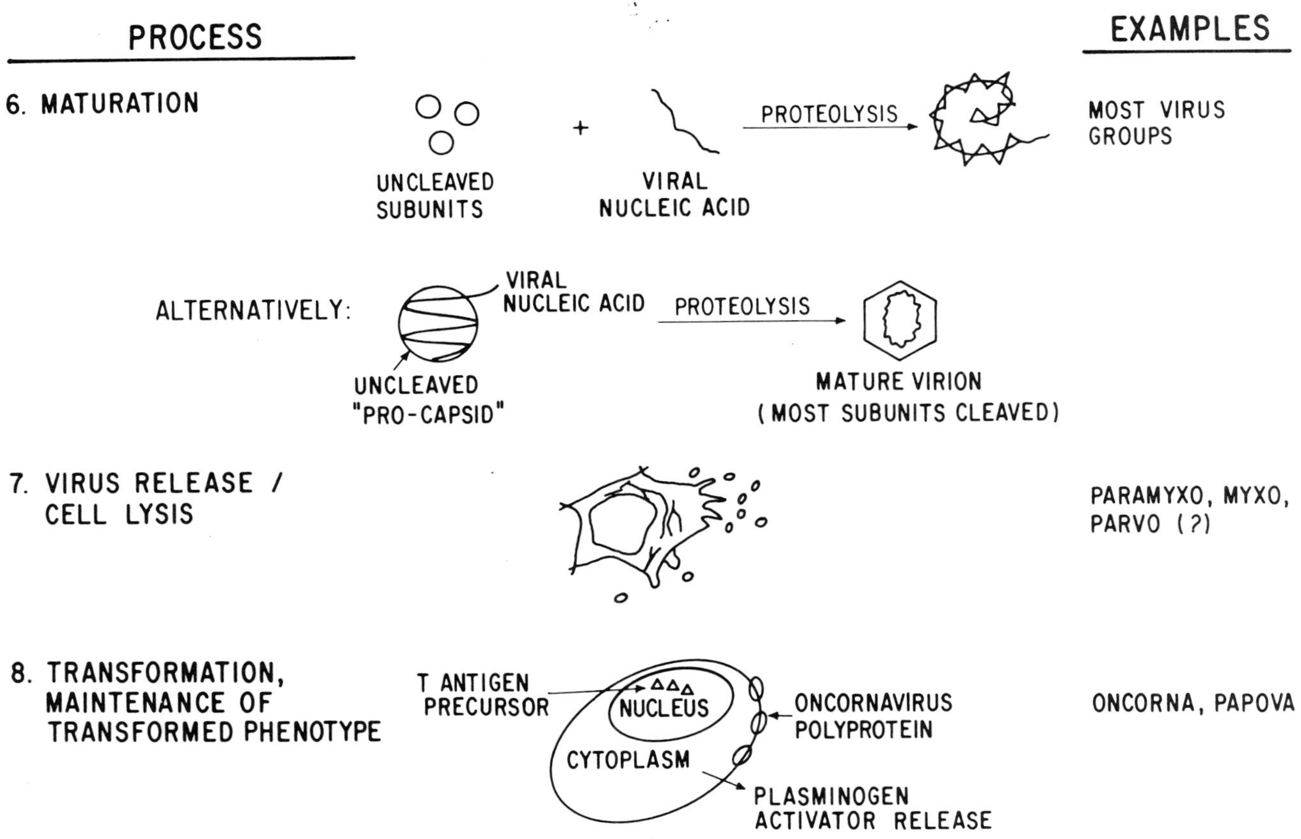

Fig. 1. Role of protein cleavage in virus replication.

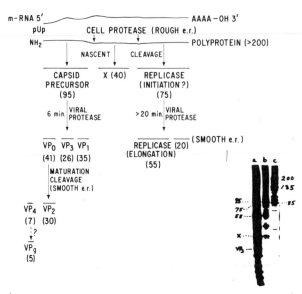

*Fig. 2. Processing of picornavirus polyprotein.
Numbers in brackets are mol. wts. in thousands. Insert:
SDS gel electrophoresis of poliovirus polypeptides.
a) Control. b) Iodoacetamide. 2) Tosyl-lysyl-chloro-
methyl ketone.*

the proteases and cleavage sites are different than
those required for the nascent cleavages.

As shown in figure 2, gel a, there are numerous
proteins present in infected cells which are not noted
in the diagram of figure 2. These have been reported,
and represent short-lived intermediates, products of
ambiguous cleavage, and artifacts of incomplete dissociation (17-19). There have been several examinations
of cell-free translation using picornavirus RNA. The
more recent studies have been successful in translating
the entire genome, and demonstrating the primary
nascent cleavages (20-23). One report found clear
evidence of production of stable capsid proteins,
implying that secondary cleavage took place (23). There
is general support from these studies of a single
"strong" initiation site (22,24), but there are also
convincing indications of a second, rarely used site
(20,21). Additional studies may help to resolve the

question. Peptide mapping techniques produced a result
implying that two initiation sites exist in vivo (25).

B. Togavirus Protein Cleavage

The group A togaviruses (alpha viruses) have been
extensively studied, and utilize single initiation
sites to synthesize polyproteins, which are then cleaved
to yield non-structural and structural end products
(for a review see 26). However, these viruses use a
more sophisticated control system than the picornaviruses. The infecting 42s genome RNA (in some reports
the RNA sediments at 49s) apparently codes for a non-
structural precursor of 200 Kd or larger, but is not
used to produce the structural polypeptides (26-30).
Instead, in the infected cells a 26s viral messenger
is present as the major species in the polysomes. This
fragment of the 42s RNA has its own unique initiation
site, and codes for a polyprotein of 130 Kd. This precursor has the small core protein (33 Kd) at its amino
terminus, and then regions which are processed proteolytically and by glycosyl transferases to yield three
envelope glycoproteins of 52, 49 and 10 Kd (27,28,31,
32).
In examining the data in most reports dealing with
the alphaviruses, a possible inconsistency with the
polyprotein model is an excess of the core polypeptide,
compared to the glycoproteins. This, however, may be
explained by termination at the carboxyl end of the
core polypeptide, degradation of the glycopeptides, or
their export from the cell. The nonstructural polyprotein is cleaved to yield four stable products, of
86, 78, 70 and 60 Kd. Genetic and biochemical studies
indicate these are involved in viral RNA synthesis and
perhaps also host cell inhibition (26,28,30,33).
The organization of the alphavirus genome is not
presently clear: if the initiation site for the non-
structural proteins is near the 5' end of the 42s RNA
(26) then the ordering of genes for alphaviruses is
rather different from the RNA phages, picorna and RNA
tumor viruses, which have structural genes at the 5'
end. If, on the other hand, the 26s RNA is derived

from the 5' end of the genome RNA, there must be an
internal initiation site in the 42s RNA, which is also
unexpected. It is obvious that the use of two classes
of m-RNA permits the alphaviruses to regulate their
RNA synthesis; the infecting RNA preferentially coding
for replicase products. The picornaviruses have adopted
an alternative control system based on selective cleavage (see below). An interesting model for the organization of a membrane-bound polyribosome has been recently proposed, based on the structural polyprotein of
Sindbis virus (34).

C. Proteolytic Regulation of Viral RNA Synthesis

    The picornaviruses and togaviruses induce RNA rep-
licases when they infect cells. The viral contribution
to the enzyme is one or more polypeptide(s), and these
are derived from precursors. The assignment of replicase function in the case of the picornaviruses is to
the proteins derived from the carboxyl third of the
polyprotein (see figure 2) (35-37). Therefore, there
is an activating role of cleavage to generate the
replicase polypeptides. In addition, replicase function
is subject to additional controls, and indirect evidence
suggests that functional lifetime of the enzyme is
regulated by limited proteolysis. The effects of protease inhibitors on viral RNA synthesis have been
studied with the conclusion that protection of the
replicase from proteolytic attack stabilizes it (1,38,
39). Time-course studies of replication of polio,
cardio and rhino-viruses have all noted selective degradation of non-structural proteins at later times after
infection (18,40-42), when viral RNA synthesis is linear
or declining. There is an inherent decay of the replicase (43) which becomes more rapid (1) in late infection.
A poliovirus mutant was described which displays a more
stable replicase than wild type virus; the mutant has
increased stability of a non-structural polypeptide
(figure 3).
    The picornaviruses are structurally similar to the
RNA bacteriophages, and their gene order is the same.
The RNA phages regulate RNA synthesis via translational

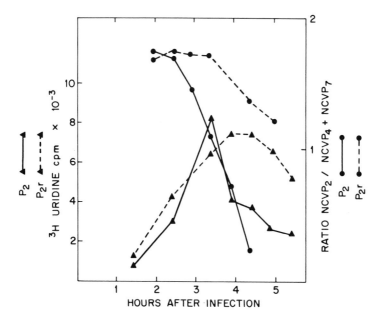

Fig. 3.   Rates of RNA synthesis and protein cleavage for poliovirus (P2) and stable mutant (P2r).

control; a control over the quantity of replicase synthesized (see figure 4 and ref. 44). The picornavirus apparently cannot accomplish this type of control and initiate protein synthesis near the amino terminus of the coat proteins. However, as indicated in figure 4, they have managed to regulate RNA synthesis by evolving specifically labile replicases, which are more rapidly cleaved as infection proceeds.

With the phages, coat protein provides a translational control of the replicase. If picornavirus capsid polypeptides possess protease activity (45,46, see below) then the coat protein may accomplish a post-translation degradation of replicase activity as a regulator of RNA synthesis.

D.   Assembly of Picornavirus Precursor Proteins

The clearest indications of a post-translational function for cleavage in virus replication comes from studies of virus assembly and maturation. In many of

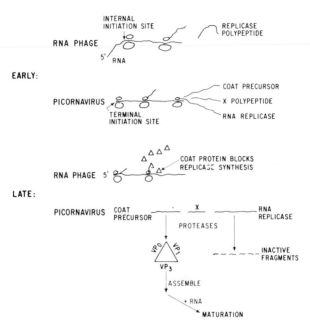

Fig. 4. Comparison of regulation of bacteriophage and picornavirus RNA synthesis, "early" and "late" in infection.

its features, the morphogenesis of picornaviruses is reminiscent of the proteolytic activations and aggregations of the complement system (reviewed in 47), although with the picornaviruses the process is entirely intracellular. Picornavirus-infected cells contain structural proteins sedimenting at 6s, 14s, 80s and 150s. The latter two structures are RNA-deficient capsids and complete virions. The smaller 6s and 14s subunits have long been known to possess different surface antigens than the fully assembled capsids, and in fact, all differ from one another antigenically (48-51). Kinetic labeling experiments have led to a summary of the assembly reaction shown in figure 5.

It is likely that the role of cleavage is to modulate conformation of the subunits as they assemble (52-55), and perhaps direct them to different subcellular structures and eventually to viral RNA. The model of assembly differs for the various picornaviruses; the cardioviruses and some rhinoviruses do not display

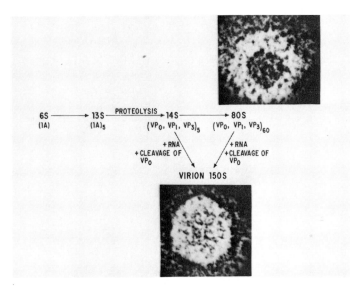

Fig. 5. Cleavage and assembly of picornavirus structural subunits. S indicates sedimentation coefficient. 1A is the 95 Kd precursor to the capsid polypeptides VP0, VP1 and VP3. (1A)₅ is a pentamer.

empty capsids. More problematic is the presence of cleaved capsid polypeptides sedimenting at 6s; these may be end products which have dissembled (54,56-58), but at present cannot be excluded from the assembly process.

There are other reports of an association of structural polyproteins and cleavage products of poliovirus with viral non-structural polypeptides, and with cellular structures, and suggestions that the association regulates viral RNA (38,53) and protein synthesis (59).

A feature common to the maturation of many animal viruses and bacteriophages (see below) is the cleavage of a structural precursor polypeptide as viral nucleic acid condenses with capsid protein. This process was first clearly described with poliovirus (57), where one or two cleavages of a 41 Kd precursor, VP0, accompany maturation. The larger VP0 fragment (~30 Kd; VP2) and a variable amount of the smaller fragment (6-10 Kd; VP4) remain in the assembled virion. This processing is apparently required for proper maturation, but does not otherwise change the surface configuration of the

capsid. As an example, rhinovirus type 2 produces uncleaved capsids which display many of the important
surface properties (antigenicity, isoelectric point,
specific attachment to host cell receptors - ref. 60)
of virions (figure 6).

Tentatively included as a cleavage product in
figure 2 is a peptide (approx. 5 Kd) which is covalently
associated with the 5' terminal RNA nucleotide of poliovirus (61,62), encephalomyocarditis virus (F. Golini,
personal communication) and rhinovirus 1A (F. Golini,
E. Wimmer, K. Lonberg-Holm and B. Korant, unpublished
results).

It is preliminary to assume that the peptide, VPg,
is derived from a larger viral peptide, since no sequence data are presently available. The peptide is
not present on viral m-RNA, but is on nascent viral RNA
in replicative intermediates (62-65); therefore, it is
removed before the RNA becomes a message. Alternatively,
it may be that those RNA strands which by chance lack
the protein are selected as message. The VPg peptides
are fairly rich in basic residues, but since there is
only one per genome, they are not sufficient to neutralize the RNA charges in packaging. The VP4 peptides,
which are internal in picornavirus capsids (66), are
better candidates for this role. The electrophoresis
of VP4 from rhinovirus type 2 and poliovirus type 2
indicates they are fairly acidic (B. Korant and K. Lonberg-Holm, unpublished results), which may actually
serve to repel rather than neutralize the RNA, in accord
with a model proposed for the condensation of T4 phage
DNA by the phage acidic internal peptides (67).

III.   RNA TUMOR VIRUS: PROTEIN CLEAVAGE AND CELL
       TRANSFORMATION

The tumor viruses have a more complicated structure
and replication cycle than do the smaller cytopathic RNA
viruses, and they also have the option of productive
infection or integration via reverse transcriptases into
stable DNA components of the host cell chromosomes.
Even in productive infection, which is the mode in which
their protein synthesis has been studied, they present
the problem of failing to significantly block cellular
protein synthesis. This has forced the use of specially

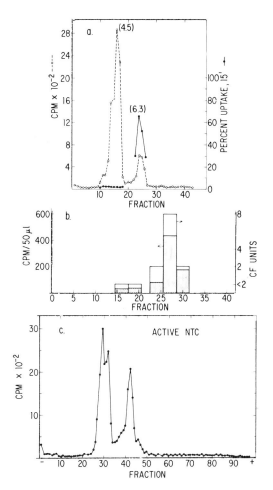

Fig. 6. Properties of rhinovirus type 2 empty cap-sids. Panel a. Isoelectric focusing gives two sub-populations, pI 6.3 and pI 4.5. Only the pI 6.3 fraction adsorbs to cells. Panel b. Complement fixa-tion by the two populations with antivirion serum. Panel c. SDS gel electrophoresis of active subpopula-tion. Composition is identical to the inactive popula-tion, and differs from virions in containing VP0, but no VP2 or VP4.

prepared, absorbed antisera to virion components in order to immunoprecipitate viral specific polypeptides. The initial reports of proteolytic processing in avian oncorna virus-infected cells (68) have been adequately confirmed and extended to murine, feline and even pri-

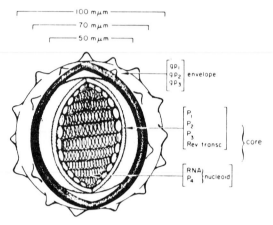

Fig. 7.   Model of an oncornavirus (84).

mate systems, and the intensity of effort has yielded
much new information about this fascinating virus group
(69-83).

Because of the use of antivirion serum, all the
proteins observed in infected cells are either present
in the virions, or are antigenically related to structural proteins. Therefore, it is necessary to briefly
review the structural features of the virions (figure 7).

The genome RNA of the oncornaviruses sediments at
60-70s, corresponding to a mol. wt. of 10×10^6, and a
potential coding capacity of 10×10^5 Kd. However, the
viral m-RNA in infected cells is not larger than 35s,
and several even smaller species are detected (85,86).
If the m-RNA is assumed to be a unique 35s molecule,
the limit of coding is reduced to about 3×10^5. The
virions consist of several polypeptides, ranging in
size from 10-80 Kd. These include the group antigens,
or core polypeptides (p30, p15, p12 and p10), the
envelope glycoproteins (gp70 and p15E), and the RNA-
dependent DNA polymerase (reverse transcriptase) of
about 80 Kd (reviewed in 73,87). The total molecular
weight of these proteins is about 250,000, which represents 80% of the information of a 35s m-RNA.

Most studies suggest the order of the viral genes
as: 5'-core proteins-reverse transcriptase-glycopro-
teins-transformation gene (sarc)-3'; but needless to say
there are numerous assumptions in the ordering process,
and some revision may yet be necessary. Satisfactory
kinetic data on processing of viral proteins may be
obtained by analyzing pulse-labeled viral antigens with

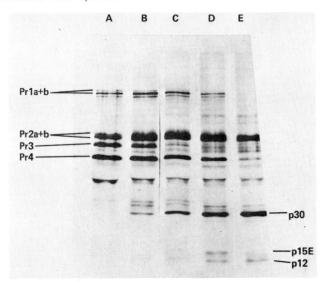

*Fig. 8. Processing of Rauscher leukemia virus
precursor proteins. SDS gel track A. pulse, B. 15
min. chase, C. 30 min., D. 60 min., E. 180 min.
p30, p15E and p12 are virion components. Molecular
weights, in Kd, of precursors are Pr2a+b, 90; Pr3, 80;
Pr4, 70. Data courtesy of R. Arlinghaus.*

standard SDS gel electrophoresis (figure 8), after
immune precipitation.

Such analyses have clearly shown the flow of label
from a polypeptide of 70-76 Kd into the group antigens
of the viral core. Cell-free translation of the genome
RNAs of Rauscher and Maloney murine leukemia viruses and
Rous sarcoma (avian) virus produced precursors which
contain the sequences of the group antigens p30 and p15
(88,89). The studies in vitro help to prove that these
proteins are truly viral gene products.

The biosynthesis of the envelope glycoprotein gp70
and p15E has been demonstrated in infected cells to
proceed through a multistage reaction requiring the
concerted actions of proteolytic enzymes and glycosyl
transferases (72-75,78-80). The use of the analog
2-deoxyglucose has indicated that some stages in the
proteolytic processing are dependent on prior additions
of carbohydrates to the substrates (73,80).

Relatively few studies of the biosynthesis of the
reverse transcriptase have appeared. The results presently indicate that precursor proteins exist for the
enzyme, and that the virion-bound form of the enzyme

may be a cleavage product of a larger intracellular form
with similar enzymatic activity (90,91).

Several laboratories have detected much larger
proteins, up to 200-300 Kd, which contain sequences of
the core and reverse transcriptase polypeptides (73,83,
92). Mainly due to an inability to show specific cleavage of these large polyproteins, there has been reluc-
tance to interpret their roles in biosynthesis. In the
opinion of some, they represent inappropriate read-
through of termination signals, or linkage of host and
viral proteins (73,83). In light of the occurrence of
viral m-RNA significantly smaller than 35s, and the
large differences in quantities of viral gene products,
it is well-advised to cautiously evaluate the roles of
very large polyproteins (see figure 9).

The importance of protein cleavage to the oncornaviruses has also been indicated in studies of condi-
tional-lethal viral mutants or non-permissive hosts.
The defect associated with certain temperature-sensitive
mutants of Rauscher murine leukemia virus (93) and an
avian sarcoma virus (94), neither of which replicate at
high temperatures, was reported to be incomplete or
inappropriate cleavages of viral precursor polypeptides.
This is reminiscent of some TS mutants of picorna and
togaviruses (sections IIA and IIB). In another case,
it was found that the basis for the lack of production
of Rous sarcoma virus particles in a transformed hamster
cell line was accumulation of the core proteins in a
76 Kd, uncleaved state (69,70). It is noteworthy that
this lack of cleavage did not prevent transformation of
the cells by the virus.

The role, if any, of the RNA tumor virus polyproteins in establishment or maintenance of the transformed
state is obviously of great interest. Unfortunately,
there is presently little information available. The
major envelope glycoprotein, gp70, is expressed on mouse
lymphocytes independent of virus production (reviewed
in 95). On the surface of murine spontaneous leukemia
cells is found a 75 Kd protein which has the amino acid
sequences of the core polypeptides of murine leukemia
virus, but differs from the usual precursor to those
polypeptides in that it contains carbohydrate. The p75
appears on the cell surfaces as a function of the age

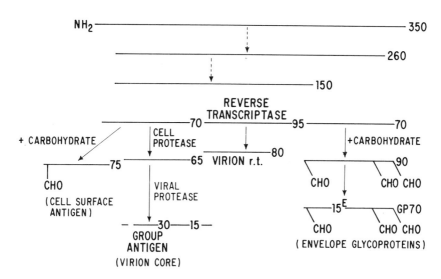

Fig. 9. Protein processing in the biosynthesis of oncornavirus structural polypeptides. Numbers indicate approximate molecular weights in thousands.

of the mouse. None is detectable at 2 mos., but a maximal amount is detected at 6 mos., just prior to the onset of leukemia. The presence of p75 on the cells does not require virus production, but correlates well with the presence of viral group antigen (95,96). The relationship of p75 to the onset of leukemia is not clear.

In some transformed avian and murine cell lines, and interestingly in some solid tumors in human patients there are substantial increases in production and release of plasminogen activators (97). These are probably proteolytic factors themselves, and can be assayed by their cleavage of plasminogen to plasmin, which activity is measured by fibrin degradation. In the avian system there is an excellent correlation between production of plasminogen activator, expression of the transformed state, and the ability of an oncornavirus to express one or more of its functions. Whether any of the viral precursors or cleavage products discussed above are involved in production of plasminogen activator, or vice versa, is unknown.

Although simian virus 40 (SV40) is a papovavirus, it is included in this section because of its ability to transform cells. The tumor, or T antigen of the virus is a non-structural polypeptide present in the nuclei of infected cells, which is intimately involved in initiation of transformation, and also in viral DNA replication in lytic infection. In a recent report (98), T antigen prepared from SV40 transformed mouse and hamster cells had a mol. wt. of 94,000, while from lytically infected monkey cells the antigen was smaller by 10 Kd. Protease inhibitor treatment of the monkey cells increased the size of the extracted T antigen to 89,000 or 94,000. The significance of the three forms of T antigen in transformation is uncertain.

IV.   ROLE OF HOST CELL PROTEASES

A.   Myxo and Paramyxoviruses: Infectivity Depends on Host Cell Proteases

The initial interaction of viruses and host cells, leading to the entry of the viral genome into the cell, is a controversial subject in terms of mechanism (for a recent review see 99). There have been several reports implicating a possible role of proteolytic reactions in early infection (45,100), but in the paramyxo- and myxovirus groups the requirement of participation of cell proteases has been most clearly demonstrated.

A photomicrograph and diagrammatic view of a representative paramyxovirus are shown in figure 10. The virions are composed of an RNA molecule surrounded by approximately 9 polypeptides and a lipid membrane (for a review see 101). Paramyxoviruses and myxoviruses have a generally similar structure, especially in terms of the surface glycoproteins which are the site of proteolytic activation of these viruses.

The replication of the myxoviruses has been of considerable clinical importance, since they include the influenza viruses. The paramyxovirus group, which includes paramyxoviruses types 1-5, measles, mumps, Sendai, and Newcastle disease virus, have been of great interest

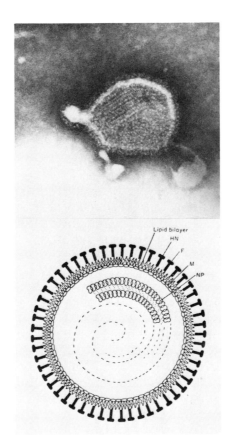

Fig. 10. Photomicrograph of Simian virus 5, negative stain x175,000. Courtesy of R. Compans. Diagram of a paramyxovirus (101). Glycoproteins HN (hemagglutinin) and F are on the surface. M is the membrane protein, within a lipid envelope. Np is nucleoprotein, found internally associated with the viral RNA.

biologically for their ability to cause persistent infections in cells of man, other mammals and avian species. The persistence is based on ability of membranes of this virus group to replicate in susceptible cells without killing them, although large quantities of virus may be produced. Pathological autoimmune reactions may then follow in an infected animal due to alterations in the surface antigens of the infected cells. The para-

myxoviruses also have the property of causing hemolysis
of erythrocytes and fusion of other cell types. The
latter has made them particularly valuable tools for
experimental cytogeneticists. Another feature of these
viruses which has been described but not clearly explained until recently are the great variations in
virulence of subtypes, and the inability to productively
infect cells of mammalian origin in culture.

Many of the biological properties mentioned for
these viruses have now been shown to be based on a
requirement for cellular proteases to activate virions,
by cleaving surface glycoproteins. The glycoproteins
are synthesized on the rough endoplasmic reticulum of
infected cells, then transferred via smooth membranes
to the plasma membrane, budding point for the new
virions (102). During this migration, cleavage of the
glycoproteins occurs, if the infected cell is to release
an active population of progeny virus.

In the case of influenza A and B, the hemagglutinin
glycoprotein can exist in two forms in the virions: as
an uncleaved polypeptide of 80 Kd, or as a cleaved di-
sulfide-bonded aggregate of 50 and 30 Kd chains (103).
For the paramyxovirus Sendai, a 65 Kd glycoprotein may
be found uncleaved, or as a 53 Kd fragment (104). Uncleaved glycoproteins are present in large amounts in
virus preparations, after which several orders of magnitude increases in infectivity, hemolysis, etc. are
observed (104-107). Activation is invariably accompanied by limited cleavage of the precursor glycoproteins. In the usual case, trypsin seems to be required
for activation (105,107-109), although other proteases
have been used to select for activation (cleavage site?)
mutants (110), which respond to chymotrypsin or elas-
tase-specific cleavages. In the case of Newcastle
disease virus, a hemagglutinin-neuraminidase glycoprotein (74 Kd) is derived from an 82 Kd precursor following cleavage by any of a variety of proteases, including
chymotrypsin, elastase and thermolysin, while tryptic
cleavage of a 68 Kd precursor yields a 56 Kd glycoprotein required in cell fusing and hemolytic activities
(111).

An important implication of the results with these
viruses comes from the comparative biology of virus

production in different cell types. The most extensively studied example is Sendai virus in avian and mammalian cells (104,108-110). Virus produced by murine or
bovine cells is of low infecticity, and lacks hemolytic
and cell fusing ability, while embryonated chicken egg-
grown virus is high in all these qualities. The difference between active and inactive preparations is the
lack of glycoprotein cleavage in the virus from low
yielding cell types.

With a virulent Newcastle disease virus, avian cells
are generally high yielders; bovine, hamster and, curi-
ously, chick embryos are non-permissive. By comparison,
virulent strains of the virus grew in all cells tested
(112). Thus, the biochemical basis for host range and
pathogenicity of this group of viruses is dependent both
on the presence of cellular proteases and the ability
(structure) of the viral substrate to be cleaved. This
model is summarized in the diagram of figure 11.

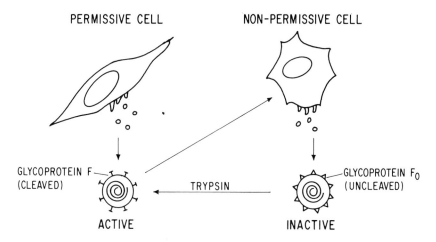

*Fig. 11. Proteolytic activation of glycoproteins
of myxo- and paramyxoviruses.*

It should be added that no study to the present has
shown directly the presence of a protease in a permissive cell which can activate the defective virus; nor
has a restrictive cell been shown to lack a crucial
cleaving enzyme. Lacking this conclusive evidence,
other alternatives remain, e.g. the permissive cell

might modify the precursor glycoprotein, so that it can
be cleaved by a protease, or the non-permissive cell
could inhibit a viral protease. There is little positive evidence for these alternative models, although a
concerted effect of a cellular plasminogen activator
and serum plasmin has been suggested in proteolytic
activation of influenza virus hemagglutinin (103,106).
     An interesting extension of this model is taken
from results with paramyxovirus type 1, strain 6/94.
This virus was isolated from fused brain tissue of
human patients suffering from multiple sclerosis (113).
The 6/94 virus is antigenically related to Sendai virus,
and will infect human brain cells or bovine kidney cells.
The progeny has very low infectivity for brain, however,
unless treated with trypsin, whereupon it can reinfect
brain cells. Once again, the difference between high
and low infectivity virus is in cleavage of a viral
glycoprotein (114). Passage of the virus through phagocytic macrophages leads to highly infectious, cleaved
progeny. It was suggested that macrophages may serve
as a (proteolytic) reservoir in the brains of MS patients, there being continuously reinfected and producing virus which infects the brain tissue (114,115).
Alternatively, elevation of serum plasmin or other
trypsin-like enzymes could serve to activate the latent
virus.

B. Cellular Proteases in Other Viral Infections

     Evidence for participation of cellular proteases
in the replication of viruses other than myxo- and
paramyxoviruses is indirect. The examples include the
following: 1) The cleavage of nascent picornavirus
polyproteins is probably carried out by cellular enzymes which are associated with polyribosomes. This
was concluded from in vivo (116) and cell-free studies
(22,116). With poliovirus polyprotein as substrate,
uninfected HeLa cell extracts were able to produce
primary cleavage products with the molecular weights
and antigens of intermediates in the normal cleavage
process. Little or no production of capsid polypeptides
was reported (116). 2) Cleavages that produce Sindbis

virus capsid polypeptides do not occur in aged (seven day old) chick embryo cells, unless actinomycin D is added. The latter treatment may have caused release of lysosomal hydrolases (117). With another togavirus, the rates of cleavage of viral precursors are different depending on the infected cell line (118,119). 3) Host restriction of RNA tumor virus growth occurs in some cell lines. The defect in mouse or hamster cells appears to lie in lack of production of viral coat protein (69,120), and the accumulation of coat protein precursors in a restrictive cell implied lack of host function required for processing (69,71). Fusion of permissive avian cells with restrictive hamster cells led to cleavage and partial rescue of a sarcoma virus (69).

There have been no detailed descriptions of cellular proteases which are utilized in the cleavage of viral proteins. In this connection, it is noteworthy that some cellular proteins including hormones and other secretory polypeptides, are produced by limited cleavage of precursors (14,15,121). The proteases involved are not fully characterized, but seem to be located on membrane-bonded organelles, or on polyribosomes (15,122, 123), have optimum activities at neutral to slightly alkaline pH, and are sensitive to phosphofluoridate or other serine esterase inhibitors (14,122).

Recent evidence points to the presence of protease activity associated with polysomes and ribosomes when cell-free extracts of uninfected cells are assayed (figure 12). Detection of the activity is facilitated by use of sensitive assays. One such assay, based on measurement of the release of radioactive cleavage fragments from insolubilized substrate is diagrammed in figure 13. This use of insolubilized substrates to quantitate purified protease has been described in detail (124).

Inhibitory treatments with chemical and protein protease inhibitors indicate the ribosome-associated enzyme is trypsin-like in HeLa cells (122). A major unanswered question is whether the protease is actually a ribosomal structural protein, or a loosely-bound activity. The protease is stable to some inhibitors of protein synthesis. However, a potent inhibitor of

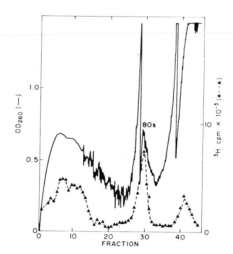

Fig. 12. Sucrose gradient sedimentation of neutral protease activity in HeLa cell extracts. The solid line indicates optical density at 260 nm, broken line the presence of protease activity. The 80s marker is the position of ribosomes. Polyribosomes are in fractions 1-15 (122).

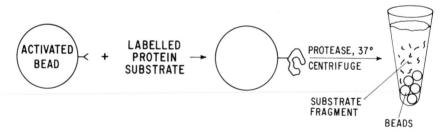

Fig. 13. Diagrammatic representation of solid phase protease assay.

the cell-free initiation of protein synthesis (125), double-stranded RNA, is likewise an inhibitor of the protease (122).

Characteristic of infection of cells by poliovirus is drastic, rapid inhibition of protein synthesis. As with treatment by double-stranded RNA, poliovirus infection depresses intracellular protease activity (39,122).

The mechanisms are probably different, however, since guanidine, which prevents viral RNA synthesis, did not stop the suppression (figure 14).

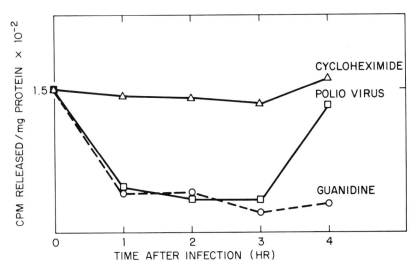

Fig. 14. Depression of HeLa cell protease activity follows poliovirus infection. The ordinate indicates protease activity, assayed by release of radioactive protein fragments from an insoluble support. Cells were infected, and treated with 100 µg/ml cycloheximide or 2 mM guanidine or untreated (39).

Indicated in section VIII.B. is the requirement for synthesis and processing of a viral protein before host protein synthesis is eliminated. The ribosome-associated protease may indicate a requirement for a proteolytic activity in synthesis of cellular proteins, which viral messages do not require. Alternatively, the virus may initially utilize the protease of native ribosomes, then replace it with another coded by the viral genome (see below). There was a recent report showing in kinetic experiments the cleavage of precursor proteins to yield two ribosomal polypeptides (126). By blocking this step, poliovirus infection could lead to ribosomes with altered structures and specificity.

V. EVIDENCE FOR VIRUS-SPECIFIC PROTEASES

In an attempt to comprehend the complex process of viral replication, it is crucial to know the origin of the enzymes which carry out the rate-limiting reactions. With viral protein cleavage, it is the enzyme(s) which produces the capsid polypeptides that is of greatest interest. The simplest possibility is that host proteases are used to cleave the virus-coded substrates. However, this reduces the ability of the virus to control the reactions, and in fact, there are several pieces of direct evidence which support the existence of a viral-coded enzyme.

Virus-infected cells contain in their cytoplasm greater quantities of protease activity than do uninfected cells (116), and the induced activity parallels the time course of infection (39,46,53,127), and the amount of virus used to initiate infection (39). Extracts of infected cells are able to carry out cleavages of viral precursors leading to production of capsid polypeptides or fragments of them (53,116,128,129,130, 131), while extracts of uninfected cells lack these activities (53,116,128,131).

There have been several reports of protease activities associated with purified virions of animal viruses, including picornaviruses, a myxovirus, a papovavirus, and the RNA tumor virus, Rous sarcoma (45,46,132, 133). Two studies have suggested individual capsid polypeptides of a cardiovirus (46) and Rous sarcoma virus (133) possess intrinsic protease activities on viral substrates. In neither of these studies was a purified substrate used, so that indirect activation of a host enzyme cannot finally be eliminated as a possible contributing factor.

The time course of induction of protease activity in poliovirus infected HeLa cells is shown in figure 15. There is decline in endogenous cellular protease activity initially. The rate of new enzyme production and the amount is positively correlated with the quantity of infecting virus (39). A characteristic of the new protease is a less alkaline optimum, compared to the major cellular protease (figure 16). However, the

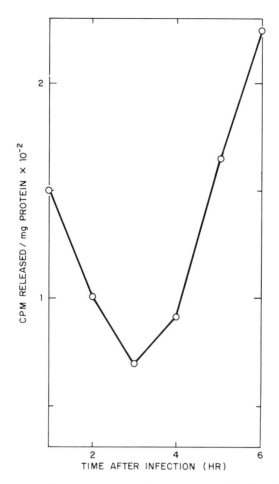

Fig. 15.   Time course of protease induction in polivirus-infected cells.   Substrate was viral precursor protein (39).

optimum pH of approximately 7 is unlike that of the major lysosomal enzymes (134,135).

Production of the new protease in poliovirus infection requires viral RNA synthesis, since the activity is not present in guanidine-treated cells (2 mM guanidine has no effect on cell-free protease assays, unpublished result). Protein synthesis is also necessary for production of the protease, because cycloheximide blocks its appearance (39). When amino acid analogs are added

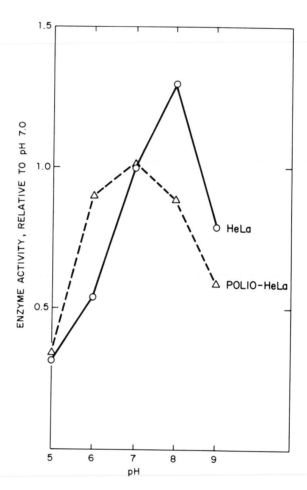

Fig. 16. Protease activity on viral precursor substrate. HeLa cells or polio-infected HeLa cells were homogenized and aliquots adjusted to various pH values. Protease activity at pH 7.0 was arbitrarily assigned a value of 1.0 (39).

to poliovirus-infected cells at midcycle of replication, virus yields are greatly diminished (136); by comparison there is a virtually normal yield of protease (figure 17). However, the protease produced is heat sensitive, compared to the enzyme from untreated cells (figure 18).

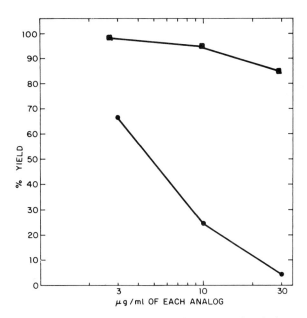

Fig. 17. HeLa cells were infected with poliovirus, and after three hours of infection, one-half the culture was treated with several levels of a mixture of aza-leucine, fluorphenylalanine and azatryptophan. After an additional three hours, the cells were harvested and extracts tested for protease activity (—■—) or assayed for infectious virus (●—●).

This indicates the enzyme contains analogs, and therefore has been synthesized at a time after infection when host protein synthesis is abolished.

A poliovirus mutant was previously described (1), which displayed a lack of regulation of viral RNA metabolism.

Extracts of cells infected with this virus display lower amounts of protease than do those of wild-type virus (figure 19). Moreover, the protease produced by the mutant virus is more temperature-sensitive than that of its parent (figure 20). These results indicate that the protease itself is coded for by the viral genome. Attempts are under way to purify the protease activity and identify it with one of the viral polypeptides.

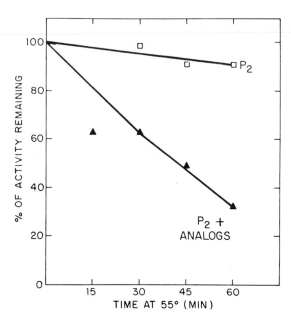

Fig. 18. Extracts from untreated and analog-treated cells (30 μg/ml of each analog) were heated at 55° for up to 60 min. All samples were then chilled, and assayed for protease activity at 35°.

VI. PROTEASES IN THE REPLICATION OF BACTERIAL VIRUSES

In depth discussion of the regulation of bacterial virus (phage) gene expression and maturation is beyond the scope of this review. However, it would be remiss not to mention briefly recent major conceptual advances in the understanding of temperate lambda phage induction and maturation of the lytic coliphage T4.

The genome of bacteriophage lambda is double-stranded DNA, which can infect E. coli, and replicate autonomously. However, the lambda DNA is commonly found physically integrated into the host chromosome. There it is regulated in part by bacterial controls, and also by a repressor protein coded by lambda, which prohibits transcription of the lytic cistrons of the viral DNA. The repressor protein has been thoroughly characterized, and antibody prepared which is specific

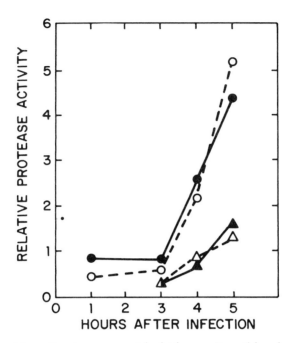

Fig. 19. Protease activities of poliovirus type 2 and a mutant, P2r, assayed in cell-free extracts. P2 protease, P2 substrate (●──●); P2 protease, P2r substrate (o----o); P2r protease, P2 substrate (▲──▲); P2r protease, P2r substrate (△----△).

to it. The antibody was used in an assay for the presence of the repressor after induction of the phage, following ultraviolet irradiation or mitomycin treatment of the host bacterium (137). Unexpectedly, it was found that concomitant with the inactivation of the repressor control, there was cleavage of the repressor protein, from an initial size of 27 Kd to a fragment of 14 Kd. The inactivation of the repressor requires the action of a set of bacterial genes which play a role also in DNA repair, filamentous growth morphology and other control mechanisms. It was suggested that these bacterial genes may code for one or more proteases, whose function it is to inactivate various repressors of repair and recombination functions (137). In support of this proposal, a tetrapeptide protease inhibitor,

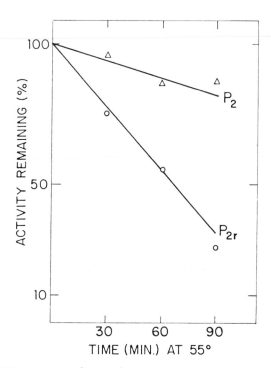

Fig. 20.   Heat inactivation of protease in extracts of P2 or P2r-infected HeLa cells.

antipain, blocked lambda phage induction, DNA repair and filament formation in E. coli, presumably by nullifying the action of specific proteases (138).

The list of classical studies of the molecular events which govern an organism must include those done with the T-even phages of E. coli. Compared to many animal viruses, these large, DNA phages are highly complex, yet there are still overall similarities in the replication processes. One of these convergent functions is the use of protein cleavage (reviewed in 67) to generate several of the capsid components during assembly and maturation (139, 140). From the virologist's standpoint, a satisfying conclusion of recent experiments is the realization that a viral gene is producing the protease.

A brief report indicated a loss of host trypsin-like function following T4 infection; replacement with

a modified form of the enzyme followed the time-course
of phage replication (141). The initial loss of the
esterase activity was subsequently confirmed; however
bacterial mutants with reduced levels of the enzyme
(protease II) permit normal phage replication (142).
Rather than relying on an endogenous or modified host
enzyme, T4 phage gene 21 codes for a protease, which
cleaves the gene 22 structural precursor to yield the
internal peptides of the mature particles.

The cleaving enzyme, or T4 phage pre-head protease,
is detectable in extracts of certain mutants blocked in
normal head assembly, but is not present in uninfected
cells, or in cells infected with phages which possess
normal capsid assembly functions (143,144). Mutants
blocked in phage gene 21, which is normally required
for maturation, have lost the protease activity in vivo
and, with certain conditional lethal mutants, in vitro
as well (145,146).

The pre-head protease is physically associated
with incomplete heads, has specificity for a structural
precursor protein (gene 22 product), and will not cleave
other phage proteins (145).

Approximately 100 molecules of protease are present
in T4 pre-heads, where they are activated by autocata-
lytic cleavage of a zymogen (147). A purification
scheme, using cleavage of a structural precursor as an
assay, gave a somewhat low yield (3-10%) of enzyme,
which contained only one polypeptide of 18.5 Kd (148).

The enzyme is not susceptible to several inhibitors
of serine, metallo, or sulfhydryl proteases. Cleavage
of soluble substrates gave appropriate N-terminal alanyl
residues and purification increased the substrate specificity of the enzyme so that it cleaved only native substrate (148). This is in contrast to crude preparations,
which could recognize both native and denatured substrates (140). The 18.5 Kd protease is approximately
95% lost from the pre-head during final morphogenesis
and probably yields small fragments which partially
remain in the phage. However, the loss of activity may
not totally be the result of digestion, since apparently
intact, but inactive protease molecules remain in the
completed heads, where they may be bound to an inhibitor
protein (147).

From detailed sequence analyses of the end groups
generated in maturation of T4 head proteins and peptides, it appears the T4 protease has specificity for
cleavage between glutamyl-alanyl residues (149,150).
However, the enzyme has additional binding features, so
that at least three residues on each side of the site
are recognized (151). Other glu-ala sites, with different residues on either side, are not cleaved in the
structural precursor, or in unrelated proteins (151).
The possible function of a second endoprotease in maturation is suggested by the presence of a lysine residue,
rather than alanine, at the new N-terminus of a cleavage product, internal peptide VII (S. Champe, personal
communication). This could be due, however, to a host
exopeptidase activity acting after the primary cleavage
event.

VII. CLEAVAGE SITES IN VIRAL PRECURSOR PROTEINS

Sequence analysis of amino and carboxyl termini of
viral proteins which are cleaved from precursors can be
used to reconstruct the primary sequence of the cleavage site. Obviously, the information obtained is only
valid if exopeptidases have not removed terminal residues after the primary peptide bond scission (this latter phenomenon has been described with proinsulin--see
ref. 14).

As indicated in the previous section, extensive
sequence analyses of cleavage sites in phage maturation
have been completed, including the primary sequence of
the precursors (150,151). It is still unclear whether
secondary or tertiary configuration of the substrate is
involved; such a conclusion will require extensive biochemical and crystallographic analyses for completeness.
With T4 phage, the glutamyl-alanyl bond is the site of
scission, but as indicated in Table 2, several residues
on either side of that site are also determinants.

In general, sequence analyses of the animal virus
precursor proteins are not available. However, there
have been determinations of N and C terminal sequences
of capsid polypeptides, which permit educated guesses of
the structure of the cleavage region (recalling the
qualification mentioned above).

TABLE II Cleavage Sites in Viral Proteins

Virus	No. of Examples	Cleavage Site Structure
T_4 Phage	1	$NH_2 \cdots\cdots\cdots H_1 X_1 Glu\S Ala\ X_2 H_2 \cdots\cdots\cdots COOH$ $\quad$ ①
		H_1, H_2 = hydrophobic (Leu,Ileu,Val)
		X_1 = Thr,Ala
		X_2 = hydrophilic (Thr,Arg,Glu)
		① = lysine, in internal peptide VII
Picornavirus	8	NH_2(blocked) $\xrightarrow{VP_4}$ LeuLeuAla$\S$AspGlnAsn(Met) $\xrightarrow{VP_2}$ LeuArgGln
capsid polypeptides		HOOC(Gln)Leu $\xrightarrow{VP_1}$ (Val)IleuGly$\S$LeuAlaGlu $\xrightarrow{VP_3}$ (Ala,Ser)Ileu(Pro)Thr(Thr)Ser
RNA Tumor Virus	9	NH_2ProLeuArg $\xrightarrow{P_{30}}$ Leu$\S$LeuAlaMet $\xrightarrow{P_{15}}$ LeuCOOH
core polypeptides		

The terminal analyses have so far been limited to picornavirus and RNA tumor virus structural polypeptides. The results are listed in Table 2.

Although the number of examples are limited, there is striking repetitiveness of carboxyl terminal leucine (152) and glutamine, or glutamine acid (153). Since both of these residues will fit the binding pocket of known proteolytic enzymes (154), it may be that the cleavage enzymes of the distinct viruses have a common origin. Alternatively, the use of helix-breaker residues may provide structural uniqueness of the region to be cleaved in the precursor (see below).

From an inspection of the accumulated data, Bachrach (154) has proposed that cleavage sites occur between α-helix maker-breaker pairs (B-bends) at exposed sites in the precursors. In support of this hypothesis are the presence of proline, glycine, serine and asparagine residues at the new amino terminii; these are all "strong" helix breakers. The new carboxyl terminal leucine and glutamine (glutamic acid) are helix formers, and might be expected to fit spacially into a single binding pocket, although clearly their chemical properties are quite distinct. The viral cleavage sites would thus be explained in a manner similar to the activation of other protein zymogens (154).

This model can be tested by additional sequencing of other viral structural and non-structural polypeptides. In addition, there have been several reports of ambiguous cleavages of picornavirus and tumor virus proteins (18,19,83). Once sequences of these altered, or transient sites are known, the requirements of the processing reactions will be better understood.

VIII.   INHIBITORS OF VIRAL PROTEOLYSIS

There are two tactics that have been widely used to inhibit viral proteolysis; these are alteration of primary structure and/or conformation of the precursors, and the inactivation of the cleaving enzymes. Some of the inhibitory treatments are particularly useful in demonstrating that proteolysis is actually occurring, and in accumulating precursor proteins for comparative peptide analyses with stable end products.

A. Alteration of the Substrate

The device most often used in the in vivo modification of the substrate is by the addition of unusual analogs of L-amino acids to the tissue culture medium. At moderate analog concentrations, viral protein synthesis continues, but the products contain the altered residues. This alters the primary sequence, and probably configuration of the precursors. In one early study, combinations of fluorophenylalanine, canavanine (arginine analog), ethionine and azetidine-2-carboxylic acid (proline analog) were used to accumulate very large poliovirus polyproteins(9). Other laboratories have used similar treatments to block proteolysis with members of the cardiovirus (155), togavirus (118), myxovirus (156) and poxvirus (157) groups. There have been contradictory reports of attempts to demonstrate RNA tumor polypeptide precursors with analogs (71,73, 75,83). This result can be easily explained in terms of loss of the antigenic determinants of the altered viral protein determinants which are required for immune precipitation (see section III). There are reports of proteolytic processing of analog-containing proteins. These reactions proceed more slowly than normal (B. Korant, unpublished results), and may signify random degradation by lysosomal enzymes.

There is another interpretation of the effect of analogs which is possible. In light of evidence for virus-induced proteases with picorna- and oncornaviruses (section V) it cannot be discounted that the analogs are altering the enzymes directly.

An argument against a direct effect of analogs on poliovirus protease comes from the experiments shown in figures 17 and 18. Protease is active in extracts of cells which had received sufficient levels of analogs to block proteolysis.

Abnormally high temperatures are known to block viral protein cleavage with numerous viruses (1,4,93, 94). Usually, a few minutes at 39-41°C is sufficient to stop the cleavage reactions, which is a temperature range well below the point of heat inactivation of the proteases (figure 20, above). Most likely the substrates are being denatured, or reversibly altered

(158), although effects on the host cells are possibly
involved.

As reviewed in previous sections, some of the precursor polypeptides of toga, oncorna and other enve-
loped viruses are glycosylated during the cleavage process. Use of inhibitors of glycosylation, e.g. 2-
deoxyglucose, glucosamine and tunicamycin have indicated
a structural requirement for glycosylation before cleavage can ensue (73,80,112). Whether the carbohydrate
causes a configurational alteration in the precursor,
or subsequent transport to a sequestered protease, is
unknown.

Zinc ion is an inhibitor of virus production (159)
and blocks protein cleavage of rhinovirus (159,160),
enterovirus (160), cardiovirus (160,161) togavirus (162)
and oncornavirus (71) precursors.

Compared to other methods of inhibition, zinc
treatment is reasonably gentle, so that there is minimal
cytotoxicity (159), and only 20-50% inhibition of protein synthesis (160), and its effect is reversible by
washing the treated cells in zinc-free medium (163).
Accumulation of large quantities of viral precursors is
possible with zinc treatment, and these may be used in
sequence analysis or as antigens.

The conclusion that zinc acts to alter the viral
substrate is based on studies of picornaviruses. The
conversion of capsid precursor is more sensitive than
the other cleavages (160,161), however increasing concentrations will accumulate progressively larger precursors (figure 21); these all contain capsid sequences
(160). With coxsackievirus (164) and rhinovirus 1A,
zinc rapidly blocks viral maturation, and zinc will
bind to the rhinovirus capsids, and prevent their crys-
tallization (163). Rhinoviruses selected for zinc resistance display altered capsid antigens (163), consistent with a binding site for zinc in the capsid sequence.
After mengovirus capsid precursor synthesis is completed,
its processing is resistant to added zinc ion (161).

A function of the picornaviruses which is poorly
understood is their ability to halt host protein synthesis. A model which has received recent support is
the simple out-competing of the cellular m-RNAs for
initiation factors by the viral RNA (165). There is

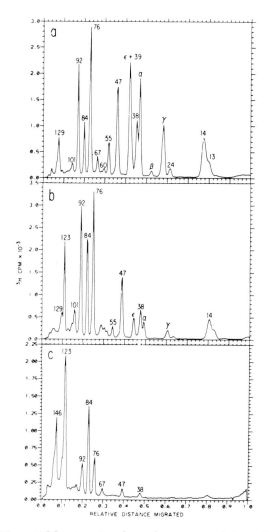

Fig. 21. Effect of zinc ions on rhinovirus type 1A protein cleavage in HeLa cells. Panel a. No treatment, b. 0.1 m$\underline{M}$ ZnCl$_2$, c. 0.8 m$\underline{M}$ ZnCl$_2$.

good evidence that this occurs in vitro (166); this, however, is not sufficient to explain many features of shut-off in infected cells (for a review see 167). More likely is the additional requirement for a viral protein in the inhibitory role. The coat protein of poliovirus has been proposed as an "equestron", modulating host ribosome function (159). If a stable viral protein is

required for inhibition, then a specific blockade of
viral protein processing should allow host protein synthesis to continue. Taking advantage of the sensitivity
of rhinovirus protein processing to the presence of zinc
ions, the requirement for cleavage of viral precursors
was assessed. HeLa cells were infected with 100 plaque-
forming units per cell of rhinovirus type 1A. Evalua-
tion of host protein synthesis was made by zonal sedimentation of cellular polysomes and by incorporation
of radioactive amino acids into cellular proteins. If
zinc ions (0.3 m$\underline{M}$) were present at the start of infection, the virus could do little harm to cellular translation (figure 22 and Table 3). In the absence of zinc,
protein synthesis was abolished (figure 22c). Zinc
alone had no effect on HeLa cell protein synthesis at
the concentration used. Viruses which are relatively
insensitive to zinc ion; poliovirus (159) and zinc-
resistant rhinovirus 1A (163) were able to stop host
protein synthesis in up to 0.3 m$\underline{M}$ zinc ion. The conclusion to be drawn from this experiment is that although
saturating levels (10^4-10^5 particles/cell) of virus were
used and viral m-RNA was introduced into the cells, host
protein synthesis was spared if viral proteins were not
synthesized and normally processed.

B.   Chemical Protease Inhibitors

     Inhibitors of well known metallo-, sulfhydryl and
serine proteases block viral proteolysis in infected
cells. Unfortunately, conclusive evidence of the nature
of the proteases involved cannot be drawn from these
studies, primarily because the inhibitors lack specificity, or possess reactive groups which can modify substrates or cellular molecules.
     The use of iodoacetamide (IAA) to stabilize polio-
virus precursors has been described (53). The precursors were subsequently cleaved in vitro to yield normal-
sized products (53), and polio RNA replicase activity
could be detected in cell-free extracts of IAA-treated
cells (38). These results could be interpreted to signify that a sulfhydryl-type protease; e.g., cathepsin B
(168), is required for some of the cleavage reactions.

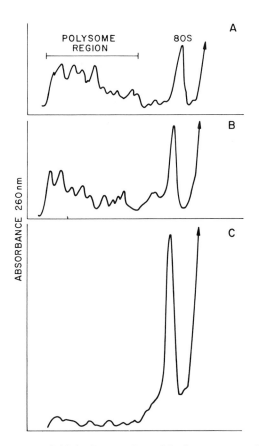

Fig. 22. Inhibition of cellular protein synthesis by rhinovirus type 1A in the presence of zinc ion. Zonal sedimentation of cell extracts in 5-46% sucrose gradients was done to display the polysomes. Extracts were made 0.5% in deoxycholate and Triton X-100 before centrifugation. Sed. right to left. Panel A. cell control, Panel B. HRV 1A-infected plus 0.3 mM Zn²⁺, Panel C. HRV 1A-infected.

However, if ^{14}C-iodoacetamide is added to infected or control cells, numerous cellular structures react, and there is substantial inhibition of protein synthesis (B. Korant, unpublished results). Energy metabolism in such cells would certainly be depressed, and ATP-requiring proteolytic reactions blocked.

TABLE III   Inhibition of HeLa Cell Protein Synthesis
by Rhinovirus Type 1A

Treatment	^{3}H amino acids cpm incorporated	% Inhibition
none	2.8×10^5	--
HRV 1A	0.6×10^5	79
HRV 1A+0.3 m$\underline{M}$ ZnCl$_2$	2.6×10^5	7

Chelators of metal ions in metallo enzymes block
processing of rhinovirus precursors (B. Korant, unpublished expts.). Ortho-phenanthroline blocks the same
cleavages as does zinc ion, moreover the two inhibitors
antagonize one another. As with IAA, phenanthroline
and EDTA disrupt protein synthesis, and cause cells to
detach from surfaces, making interpretation of results
in proteolysis experiments difficult.

The esterase inhibitor diisopropyl phosphofluoridate (DFP) was shown to block the cleavages of poliovirus (9) and a myxovirus (169). This implicates proteases with serine-containing active sites.

The most interesting class of protease inhibitors
in viral studies has been the chloromethyl ketone derivatives of amino acids (reviewed in 170,171). These
were designed as affinity labels of serine-type proteases, and react irreversibly with histidine and serine
residues in the active sites of proteases. There is a
basis for selectivity of the chloromethyl ketones:
knowing the substrate specificity of two distinct proteases, Shaw and colleagues synthesized phenylalanyl and
lysyl derivatives with specificity for chymotrypsin and
trypsin, respectively. This specificity led to studies
on inhibition of poliovirus protein cleavage (116) and
subsequently to other virus groups, often with positive
results (reviewed in 1,4).

There have been numerous conflicting reports on the
efficacy of particular chloromethyl ketones. With picornaviruses, this can perhaps be explained by the various cell types used supplying distinct cleavage enzymes
(116). This should result in host-dependent differences
in cleavage, and there has been some biochemical evidence

for this (116,172). There has been a substantial number of reports on effects of chloroketones on uninfected
cell morphology and metabolism (reviewed in 171). A
straightforward interpretation implies that serine proteases do much to regulate cellular processes, and this
may be correct. However, there are several troublesome
facts to be fit into any model of chloroketone action.
For example, cleavage of poliovirus precursors was
blocked by a D-rotatory chloroketone of phenylalanine
(173). If the enzyme were actually chymotrypsin-like,
the D-inhibitor should have been ineffectual. The various chloroketones, particularly tosylphenylalanine
chloromethyl ketone (TPCK) block protein synthesis
(174); this has been attributed to reactivity with glutathione (175), or to induction by TPCK of an inhibitory
substance (G. Koch, personal communication). The property of non-specific alkylation of sulfhydryl groups
and non-active site histidines and serine is difficult
to deal with at present, but may be of relevance. For
example, it was proposed that TPCK directly alters the
antigenic specificity of a papova virus precursor (98),
implying a chemical or conformation change, which con-
ceivably could affect normal cleavage.

     In summary, chemical inhibitors of cleavage are
available, but none appears to be satisfactorily directed against either a viral substrate or a specific
protease. Additional experimentation will be required,
if eventually an inhibitor of proteolysis is to develop
into a clinically useful antiviral agent.

IX. ACKNOWLEDGMENTS

     My thanks go to the following for supplying infor-
mation for this review, often prior to publication:
Drs. R. Arlinghaus, H. Bachrach, B. Butterworth, S.
Champe, L. Caliguiri, R. Eisenman, F. Golini, G. Koch,
H. Koprowski, D. Scraba, M. Showe, D. Tershak, G. Vande
Woude and E. Wimmer. Special appreciation goes to R.
Arlinghaus, R. Compans and P. Choppin for permitting
publication of their data. My colleagues B. Butter-
worth, N. Chow, J. Kauer, E. Knight, K. Lonberg-Holm
and F. Yin participated in the studies reported.

R.Z. Lockart provided advice and support, A. Vilda and
M. Parise performed excellent technical assistance and
S. Kelly prepared the manuscript.

X. REFERENCES

1. Korant, B.D., in "Proteases and Biological Control"
    (E. Reich, D. Rifkin, and E. Shaw, Eds.), p. 621,
    Cold Spring Harbor Lab., 1975.
2. Showe, M., and Kellenberger, E. in "Control Pro-
    cesses in Viral Multiplication" (D. Burke and W.
    Russell, Eds.), p. 407, Cambridge Univ. Press,
    London, 1975.
3. Hershko, A., and Fry, M., Ann. Rev. Biochem. 44,
    775 (1975).
4. Butterworth, B., in "Current Topics in Microb. and
    Immunol.", Springer-Verlag, New York, in press.
5. Comprehensive Virology (H. Fraenkel-Conrat and R.
    Wagner, Eds.) Vol. 1-7, Plenum Press, New York,
    1975.
6. Summers, D., and Maizel, J., Proc. Nat. Acad. Sci.
    U.S. 59, 966 (1968).
7. Holland, J., and Kiehn, E., Proc. Nat. Acad. Sci.
    U.S. 60, 1015 (1968).
8. Jacobson, M., and Baltimore, D., Proc. Nat. Acad.
    Sci. U.S. 61, 77 (1968).
9. Jacobson, M., Asso, J., and Baltimore, D., J. Mo-
    lec. Biol. 49, 657 (1970).
10. Shatkin, A., Ann. Rev. Biochem. 43, 643 (1974).
11. Butterworth, B., Hall, L., Stoltzfus, C., and
    Rueckert, R., Proc. Nat. Acad. Sci. U.S. 68, 3083
    (1971).
12. Taber, R., Rekosh, D., and Baltimore, D., J. Virol.
    8, 395 (1971).
13. Summers, D., and Maizel, J., Proc. Nat. Acad. Sci.
    U.S. 68, 2852 (1971).
14. Steiner, D., Kemmler, W., Tager, H., Rubenstein,
    A., Lernmark, A., and Zuhlke, H., in "Proteases
    and Biological Control," (E. Reich, D. Rifkin, and
    E. Shaw, Eds.), p. 531, Cold Spring Harbor Lab.,
    1975.
15. Devillers-Thiery, A., Kindt, T., Scheele, G., and

Blobel, G., Proc. Nat. Acad. Sci. U.S. 72,5016 (1975).

16. Butterworth, B., and Rueckert, R., Virology 50, 535 (1972).

17. Abraham, G., and Cooper, P., J. Gen. Virol. 29, 199 (1975).

18. McClean, C., Mathews, T., and Rueckert, R., J. Virol. 19, 903 (1976).

19. Beckman, L., Caliguiri, L., and Lilly, L., Virology 73, 216 (1976).

20. Oberg, B., and Shatkin, A., Proc. Nat. Acad. Sci. U.S. 69, 761 (1975).

21. Celma, M., and Ehrenfeld, E., J. Molec. Biol. 98, 761 (1975).

22. Villa-Komaroff, L., Guttman, N., Baltimore, D., and Lodish, H., Proc. Nat. Acad. Sci. U.S. 72, 4157 (1975).

23. Chattarjee, N., Polatnik, J., and Bachrach, H., J. Gen. Virol. 32, 383 (1976).

24. Smith, A., Eur. J. Biochem. 33, 301 (1973).

25. Abraham, G., and Cooper, P., J. Gen. Virol. 29, 215 (1975).

26. Lachmi, B., and Kaariainen, L., Proc. Nat. Acad. Sci. U.S. 73, 1936 (1976).

27. Simmons, D., and Strauss, J., J. Molec. Biol. 86, 397 (1974).

28. Lachmi, B., Glanville, N., Keranen, S., and Kaariainen, L., J. Virol. 16, 1615 (1975).

29. Kaluza, G., J. Virol. 19,1 (1976).

30. Bracha, M., Leone, A., and Schlesinger, M., J. Virol. 20, 612 (1976).

31. Cancedda, R., and Schlesinger, M., Proc. Nat. Acad. Sci. U.S. 71, 1843 (1974).

32. Clegg, J., Nature (London) 254, 454 (1975).

33. Pfferkorn, E., and Shapiro, D., in "Comprehensive Virology," (H. Fraenkel-Conrat and R. Wagner, Eds.) Vol. 2, p. 171. Plenum Press, New York, 1974.

34. Wirth, D., Katz, F., Small, B., and Lodish, H., Cell 10, 253 (1977).

35. Cooper, P., Geissler, E., Scotti, P., and Tannock, G., in "Strategy of the Viral Genome," (G. Wolstenholme and M. O'Connor, Eds.), p. 75. Williams & Wilkins, Baltimore, 1971.

36. Lundquist, R., Ehrenfeld, E., and Maizel, J., Proc.
     Nat. Acad. Sci. U.S. 71, 4773 (1974).
37. Traub, A., Diskin, B., Rosenberg, H., and Kalmar,
     E., J. Virol. 18, 375 (1976).
38. Roder, A., and Koschel, K., J. Virol. 14, 846
     (1974).
39. Korant, B., in "In Vitro Transcription and Trans-
     lation of Viral Genomes," (A. Haenni and G. Beaud,
     Eds.), p. 273. INSERM, Paris, 1975.
40. Lucas-Lenard, J., J. Virol. 14, 261 (1974).
41. Paucha, E., Seehafer, J., and Colter, J., Virology
     61, 315 (1974).
42. Paucha, E., and Colter, J., Virology 67, 300
     (1975).
43. Baltimore, D., in "The Biochemistry of Viruses,"
     (H. Levy, Ed.), p. 101. Dekker, New York, 1969.
44. Sugiyama, T., Korant, B., and Lonberg-Holm, K.,
     Ann. Rev. Microb. 26, 467 (1972).
45. Holland, J., Doyle, M., Perrault, J., Kingsbury,
     D., and Etchison, J., Bioch. Biophys. Res. Commun.
     46, 634 (1972).
46. Lawrence, C., and Thach, R., J. Virol. 15, 918
     (1975).
47. Müller-Eberhard, H., Ann. Rev. Biochem. 44, 799
     (1975).
48. Polson, A., Selzer, G., and van den Ende, M.,
     Biochem. Biophys. Acta 24, 600 (1957).
49. Watanabe, Y., Watanabe, K., and Hinuma, Y., Bio-
     chem. Biophys. Acta 61, 976 (1962).
50. Scharff, M., Maizel, J., and Levintow, L., Proc.
     Nat. Acad. Sci. U.S. 51, 329 (1964).
51. Ghendon, Y., and Yakobson, E., J. Virol. 8, 589
     (1971).
52. Perlin, M., and Phillips, B., Virol. 53, 107
     (1973).
53. Korant, B., J. Virol. 12, 556 (1973).
54. McGregor, S., Hall, L., and Rueckert, R., J. Virol.
     15, 1107 (1975).
55. McGregor, S., and Rueckert, R., J. Virol. 21, 548
     (1977).
56. Jacobson, M., and Baltimore, D., J. Molec. Biol.
     33, 363 (1968).
57. Talbot, P., and Brown, F., J. Gen. Virol. 15, 163

(1972).

58. Butterworth, B., Grunert, R., Korant, B., Lonberg-
     Holm, K., and Yin, F., Archiv. Virol. 51, 169
     (1976).
59. Cooper, P., Steiner-Pryor, A., and Wright, P.,
     Intervirol. 1, 1 (1973).
60. Korant, B., Lonberg-Holm, K., Yin, F., and Noble-
     Harvey, J., Virology 63, 384 (1975).
61. Lee, Y., Nomoto, A., Detjen, B., and Wimmer, E.,
     Proc. Nat. Acad. Sci. U.S. 74, 59 (1977).
62. Flanegan, J., Pettersson, R., Ambros, V., Hewlett,
     M., and Baltimore, D., Proc. Nat. Acad. Sci. U.S.
     74, 961 (1977).
63. Hewlett, M., Rose, J., and Baltimore, D., Proc.
     Nat. Acad. Sci. U.S. 73, 327 (1976).
64. Nomoto, A., Lee, Y., and Wimmer, E., Proc. Nat.
     Acad. Sci. U.S. 73, 375 (1976).
65. Fernandez-Munoz, R., and Darnell, J., J. Virol.
     18, 719 (1976).
66. Lonberg-Holm, K., and Butterworth, B., Virology
     71, 207 (1976).
67. Laemlli, U., Paulson, J., and Hitchins, V., J.
     Supram. Struct. 2, 276 (1974).
68. Vogt, V., and Eisenman, R., Proc. Nat. Acad. Sci.
     U.S. 70, 1734 (1973).
69. Eisenman, R., Vogt, V., and Diggelman, H., Cold
     Spr. Hbr. Symp. Quan. Biol. 39, 1067 (1965).
70. Oskarsson, M., Robey, W., Harris, C., Fischinger,
     P., Haapala, D., and Vande Woude, G., Proc. Nat.
     Acad. Sci. U.S. 72, 2380 (1975).
71. Vogt, V., Eisenman, R., and Diggelman, H., J.
     Molec. Biol. 96, 471 (1975).
72. Dickson, C., Puma, J., and Nandi, S., J. Virol.
     17, 275 (1976).
73. Shapiro, S., Strand, M., and August, J., J. Molec.
     Biol. 107, 459 (1976).
74. Forschammer, J., Nexo, B., and Vaughan, M., Viro-
     logy 71, 124 (1976).
75. Van Zaane, D., Dekker-Michielsen, M., and Bloemers,
     A., Virology 75, 113 (1976).
76. Janjoom, G., Naso, R., and Arlinghaus, R., J. Vi-
     rol. 19, 1054 (1976).
77. Okasinski, G., and Velicer, L., J. Virol. 20, 96

(1976).

78. Arcement, L., Karshin, W., Naso, R., Janjoom, G., and Arlinghaus, R., Virology 69, 763 (1976).

79. Naso, R., Arcement, L., Karshin, W., Janjoom, G., and Arlinghaus, R., Proc. Nat. Acad. Sci. U.S. 73, 2326 (1976).

80. Famulari, N., Buchhagen, D., Klenk, H., and Fleissner, E., J. Virol. 20, 501 (1976).

81. Robey, W., Oskarsson, M., Vande Woude, G., Naso, R., Arlinghaus, R., Haapalo, D., and Fischinger, P., Cell 10, 79 (1977).

82. Tsiapalis, C., Nature (London) 266, 27 (1977).

83. Janjoom, G., Naso, R., and Arlinghaus, R., Virology in press.

84. Bader, J., in "Comprehensive Virology," (H. Fraenkel-Conrat and R. Wagner, Eds.), Vol. 4, p. 253. Plenum Press, New York, 1975.

85. Fan, H., and Baltimore, D., J. Molec. Biol. 80, 93 (1973).

86. Manzel, W., Delius, H., and Duesberg, P., Proc. Nat. Acad. Sci. U.S. 71, 4541 (1974).

87. Stromberg, K., Hurley, N., Davis, N., Rueckert, R., and Fleissner, E., J. Virol. 13, 513 (1974).

88. Kerr, I., Olshevsky, U., Lodish, H., and Baltimore, D., J. Virol. 18, 627 (1976).

89. Pawson, T., Martin, G., and Smith, A., J. Virol. 19, 950 (1976).

90. Gerwin, B., Smith, S., and Peebles, P., Cell 6, 45 (1975).

91. Moelling, K., Cold Spr. Hbr. Symp. Quan. Biol. 39, 969 (1974).

92. Van Zaane, D., Gielkens, A., Dekker-Michielson, M., and Bloemers, H., Virology 67, 594 (1975).

93. Stephenson, J., Tronick, S., and Aaronson, S., Cell 6, 543 (1975).

94. Rohrschneider, J., Diggelman, H., Ogura, H., Friis, R., and Bauer, H., Virology 75, 177 (1976).

95. Tung, J., Yoshiki, T., and Fleissner, E., Cell 9, 573 (1976).

96. Ledbetter, J., Nowinski, R., and Emery, S., J. Virol. 22, 65 (1977).

97. Rifkin, D., Beal, L., and Reich, E., in "Proteases and Biological Control," (E. Reich, D. Rifkin, and

E. Shaw, Eds.), p. 841. Cold Spring Harbor Lab.,
1975.

98. Carroll, R., and Smith, A., <u>Proc. Nat. Acad. Sci.
U.S.</u> 73, 2254 (1976).

99. Lonberg-Holm, K., and Phillipson, L. in "Monographs in Virology," (J. Melnick, Ed.), Vol. 9,
p. 1. Karger, Basel, 1974.

100. Habermehl, K., Diefenthal, W., and Buchholz, in
"Adv. Biosciences," (G. Raspe, Ed.), Vol. 11, p.
41. Pergamon Press, Vieweg, 1973.

101. Choppin, P., and Compans, R., in "Comprehensive
Virology," (H. Fraenkel-Conrat and R. Wagner,
Eds.), Vol. 4, p. 94. Plenum Press, New York,
1975.

102. Nagai, Y., Ogura, H., and Klenk, H., <u>Virology</u> 69,
523 (1976).

103. Scheid, A., and Choppin, P., <u>Virology</u> 57, 475
(1974).

104. Klenk, H., Rott, R., Orlich, M., and Blodorn, N.,
<u>Virology</u> 68, 426 (1975).

105. Lazarowitz, S., and Choppin, P., <u>Virology</u> 68, 440
(1975).

106. Ohuchi, M., and Homma, M., <u>J. Virol.</u> 18, 1147
(1976).

107. Homma, M., <u>J. Virol.</u> 8, 619 (1971).

108. Homma, M., and Ohuchi, M., <u>J. Virol.</u> 12, 1457
(1973).

109. Scheid, A., and Choppin, P., <u>Virology</u> 69, 265
(1976).

110. Nagai, Y., and Klenk, H., <u>Virology</u> 77, 125 (1977).

111. Nagai, Y., Klenk, H., and Rott, R., <u>Virology</u> 72,
494 (1976).

112. Ter Meulen, V., Koprowski, H., Iwasaki, Y., Kack-
ell, Y., and Muller, D., <u>Lancet</u> 2, 1 (1972).

113. Waters, D., Koprowski, H., and Lewandowski, L.,
<u>Med. Microb. Immunol.</u> 160, 235 (1974).

114. Waters, D., Koprowski, H., and Lewandowski, L.,
<u>J. Neur. Sci.</u> 25, 491 (1975).

115. Korant, B., <u>J. Virol.</u> 10, 751 (1972).

116. Snyder, H., and Sreevalsan, T., <u>Bioch. Biophys.
Res. Commun.</u> 54, 24 (1973).

117. Snyder, H., and Sreevalsan, T., <u>J. Virol.</u> 13,
541 (1974).

118. Morser, M., and Burke, D., J. Gen. Virol. 22, 395 (1974).
119. Ray, U., Soeiro, R. and Fields, B., J. Virol. 18, 370 (1976).
120. Neurath, H., and Walsh, K., Proc. Nat. Acad. Sci. U.S. 73, 3825 (1976).
121. Korant, B., Bioch. Biophys. Res. Commun. 74, 926 (1977).
122. Dobberstein, B., Blobel, G., and Chua, N., Proc. Nat. Acad. Sci. U.S. 74, 1082 (1977).
123. Siever, E., Analyt. Bioch. 74, 592 (1976).
124. Hunt, T., and Ehrenfeld, E., Nature New Biology 230, 91 (1971).
125. Mackie, G., FEBS Letters 71, 178 (1976).
126. Kiehn, E., and Holland, J., J. Virol. 5, 358 (1970).
127. Esteban, M., and Kerr, I., Eur. J. Bioch. 45, 567 (1974).
128. Clegg, C., and Kennedy, I., Eur. J. Bioch. 53, 175 (1975).
129. Tattersall, P., Cawte, P., Shatkin, A., and Ward, D., J. Virol. 20, 273 (1976).
130. Everitt, E., Meador, S., and Levine, A., J. Virol. 21, 199 (1977).
131. Friedmann, T., J. Virol. 20, 520 (1976).
132. Von der Kelm, K., Proc. Nat. Acad. Sci. U.S. 74, 911 (1977).
133. Bohley, P., Kirscke, H., Langner, J., Ansorge, S., Wiederanders, B., and Hanson, H., in "Tissue Proteinases," (A. Barrett and J. Dingle, Eds.), p. 187. Elsevier, New York, 1971.
134. Koschel, K., Aus, H., and ter Meulen, V., J. Gen. Virol. 25, 359 (1974).
135. Wecker, E., and Schonne, E., Proc. Nat. Acad. Sci. U.S. 47, 278 (1961).
136. Roberts, J., and Roberts, C., Proc. Nat. Acad. Sci. U.S. 72, 147 (1975).
137. Meyn, M., Rossman, T., and Troll, W., Proc. Nat. Acad. Sci. U.S. 74, 1152 (1977).
138. Eddleman, H., and Champe, S., Virology 30, 471 (1966).
139. Kurtz, M., and Champe, S., J. Virology, in press.
140. Poglozov, B., and Levshenko, M., J. Molec. Biol.

84, 463 (1974).
141. Choy, W., and Champe, S., Virology, in press.
142. Goldstein, J., and Champe, S., J. Virol. 13, 419
     (1974).
143. Bachrach, U., and Beuchetrit, L., Virology 59,
     51 (1974).
144. Onorato, L., and Showe, M., J. Molec. Biol. 92,
     395 (1975).
145. Giri, J., McCullough, J., and Champe, S., J.
     Virol. 18, 894 (1976).
146. Showe, M., Isobe, E., and Onorato, L., J. Molec.
     Biol. 107, 55 (1976).
147. Showe, M., Isobe, E., and Onorato, L., J. Molec.
     Biol. 107, 35 (1976).
148. Isobe, T., Black, L., and Tsugita, A., J. Molec.
     Biol. 102, 349 (1976).
149. Tsugita, A., Black, L., and Showe, M., J. Molec.
     Biol. 98, 271 (1975).
150. Isobe, T., Black, L., and Tsugita, A., Proc. Nat.
     Acad. Sci. U.S. 73, 4205 (1976).
151. Oroszlan, S., Summers, M., Foreman, C., and
     Gilden, R., J. Virol. 14, 1559 (1974).
152. Ziola, B., and Scraba, D., Virology 71, 111
     (1974).
153. Bachrach, H., in "Virology in Agriculture," (J.
     Romberger, Ed.), p. 3. Allanheld, Osmun, Mont-
     clair, 1977.
154. Dobos, P., and Martin, E., J. Gen. Virol. 17, 197
     (1972).
155. Etchison, J., Doyle, M., Penhoet, E., and Holland,
     J., J. Virol. 7, 155 (1971).
156. Katz, E., and Moss, B., J. Virol. 6, 717 (1970).
157. Garfinkle, B., and Tershak, D., J. Molec. Biol.
     59, 537 (1971).
158. Korant, B., Kauer, J., and Butterworth, B.,
     Nature (London) 248, 588 (1974).
159. Butterworth, B., and Korant, B., J. Virol. 14,
     282 (1974).
160. Nakai, K, and Lucas-Lenard, J., J. Virol. 18, 918
     (1976).
161. Bracha, M., and Schlesinger, M., Virology 72, 272
     (1976).
162. Korant, B., and Butterworth, B., J. Virol. 18,

298 (1976).

163. Gevaudau, P., Pieroni, G., Gevaudau, M., and Charrel, J., Ann. Inst. Pasteur 122, 115 (1972).

164. Lawrence, C., and Thach, R., J. Virol. 14, 598 (1974).

165. Golini, F., Thach, S., Birge, C., Safer, B., Merrick, W., and Thach, R., Proc. Nat. Acad. Sci. U.S. 73, 3040 (1976).

166. Carrasco, L., and Smith, A., Nature (London) 264, 807 (1976).

167. Snellman, O., in "Tissue Proteinases," (A. Barrett and J. Dingle, Eds.), p. 29. Elsevier, New York, 1971.

168. Klenk, H., and Rott, R., J. Virol. 11, 823 (1973).

169. Shaw, E., in "Proteases and Biological Control," (E. Reich, D. Rifkin, and E. Shaw, Eds.), p. 455. Cold Spring Harbor Lab., 1975.

170. Powers, J., in "Chemistry and Biochemistry of Amino Acids, Peptides and Proteins," (B. Weinstein, Ed.), Vol. 4. Dekker, New York, in press.

171. Lewenton-Kriss, S., and Mandel, B., Virology 48, 666 (1972).

172. Summers, D., Shaw, E., Stewart, M., and Maizel, J., J. Virol. 10, 880 (1972).

173. Pong, S., Nuss, D., and Koch, G., J. Biol. Chem. 250, 240 (1975).

174. Freedman, M., Friedberg, D., Mucha, J., and Troll, W., Bioch. Pharm. 22, 2441 (1973).

HUMAN NEUTROPHIL ELASTASE AND THE PROTEASE-PATHOGENESIS MODEL OF PULMONARY EMPHYSEMA[1]

Aaron Janoff[2]

Health Sciences Center, State University of New York at Stony Brook

Elastolytic proteases from leukocytes and alveolar macrophages have been implicated as pathological determinants of pulmonary emphysema. In this paper, we review our work on the purification and characterization of an elastase from human neutrophilic leukocytes and on some experiments pointing to its role in this disease. Neutrophil elastase can be purified to over 95 percent electrophoretic homogeneity by affinity chromatography against phenyl-butylamine-Sepharose. The purified enzyme has a MW of 34,400; it hydrolyzes typical synthetic substrates of porcine

[1]*Supported by a grant from the National Heart, Lung and Blood Institute (HL-14262).*
[2]*The author gratefully acknowledges the collaboration of the following individuals in different phases of the work reviewed in this article: Dr. Gad Feinstein, Tel Aviv University, Israel. Drs. Bruce Sloan, George Weinbaum, Victor Damiano and Philip Kimbel, Albert Einstein Medical Center, Philadelphia, Pa. Drs. Jules Elias, Philip Kane and Robert A. Sandhaus, State University of New York at Stony Brook. Marie Blue, Harvey Carp, Dorothy Lee and Rosemarie Dearing, State University of New York at Stony Brook.*

*elastase; and it gives a single precipitin line
after immunodiffusion against its specific antiserum.
Human neutrophil elastase degrades dog lung elastin
in vitro and can produce emphysema-like lesions when
instilled into dog lung in vivo. Staining with
specific antibody reveals the leukocyte enzyme bound
to alveolar elastic fibers in the lesion area.
Using similar staining methods, one can also demonstrate neutrophil elastase in freshly resected human
lungs, suggesting that the enzyme may attack alveolar elastic fibers in man. In normal lung, this
attack is minimized by endogenous protease inhibitors. However, we observed that cigarette smoke
condensate prevents elastase inhibition by lung anti-
proteases in vitro. Thus causation of emphysema by
proteolysis may be exacerbated in cigarette smokers
by defective protease inhibition.*

I. INTRODUCTION

Pulmonary emphysema is one of three chronic
obstructive lung diseases in man, the other two being
chronic bronchitis and asthma. In the case of emphysema, the major anatomic change in the lungs is a
permanent hyper-inflation of alveoli and respiratory
bronchioles, accompanied by destruction of their walls.
The obstructive component of the disease is due to the
premature collapse of fine conducting airways (1-2mm in
diameter) during expiration. This functional abnormality is thought to result directly from the major anatomic change; namely, the destruction of alveolar septa.
Because the finest conducting airways of the lung are
devoid of supporting cartilage plates, they depend on a
"guy-wire" arrangement of supporting connective tissue
fibers in order to remain open in the face of the
increased compressive force of expiration. The connective tissue fibers or "guy-wires" are components of the
surrounding alveolar walls. Thus, destruction of the
latter renders the fine bronchioles traversing emphysematous lung tissue highly susceptible to collapse.
        Currently, the most widely accepted explanation of
the alveolar wall destruction seen in pulmonary emphy-

sema is the so-called protease-pathogenesis hypothesis
(1). According to this theory, the disease process is
initiated by conditions which give rise to an excess of
free protease activity in the lung over the level of
protease inhibitory activity in that organ. Two major
lines of evidence lend support to this view. The first
is that emphysema can be produced experimentally by
instillation of proteolytic enzymes into the airways
(2). In these animal models of emphysema, protease
excess appears to result directly from experimental
overload of the tissues with exogenous enzymes. The
second line of support comes from the clinical observation that an inherited form of the disease is positively correlated with genetically transmissable deficiency
in alpha 1-proteinase inhibitor, the major antiprotease
of serum (3). Thus, in human emphysema associated with
genetic deficiency of this inhibitor, protease excess
in the lung could result from lower than normal levels
of a major antiprotease throughout the body. The
excess lung protease activity, initiated by either of
these two mechanisms (exogenous enzyme overload or
inhibitor deficiency), can lead to unimpeded digestion
of connective tissue components and, finally, to loss
of alveolar septa. With progressive septal loss, the
anatomic and physiologic stigmata described above
become apparent; namely, permanent enlargement of
respiratory airspaces and obstructive collapse of small
airways during expiration.

Cigarette smoking is recognized to be a major risk
factor associated with the development of pulmonary
emphysema in man (4), but very little is known about
the mechanisms by which cigarette smoking might favor
excess of free proteases in the lung. One possibility
is that smoking affects the normal quantity, distribution or activity of lung antiproteases. We shall
return to this question later.

Several recent developments have suggested that
lung elastic fibers are the critical targets of unre-
strained protease activity in the pathogenesis of
emphysema. At the same time, proteases with elastase
activity are thought to be the chief enzymes responsible for the development of the emphysematous changes.
Some of the evidence in support of these statements can
be summarized as follows. First, only proteases with

elastase activity can produce emphysema in experimental
animals (2), and the severity of the disease produced
is directly correlated with the elastolytic titer of
the instilled enzyme (5,6). Second, chemical studies
of lungs in animal models of emphysema point to early
reductions in tissue elastin content (7), while
electron microscopic studies of these same lungs reveal
major abnormalities in alveolar elastin structure, at a
later stage in the disease process (7). Moreover, pretreatment of experimental animals with a lathyrogen,
beta-amino-proprionitrile, which is known to inhibit
cross-link formation in newly synthesized elastin,
greatly increases the severity of the lung lesions
induced by instillation of an elastolytic protease (8).
   At least four kinds of cells in the human body are
now known to contain an elastase or to be capable of
secreting proteases with elastolytic activity, when
appropriately stimulated. These are: neutrophilic
leukocytes (9-17), alveolar macrophages of smokers (18),
platelets (19) and pancreatic alpha cells (20). At
present, the first two are regarded as especially
likely sources of lung-damaging elastases in emphysema.
In the case of the alveolar macrophage, its role in
emphysema was suggested by the observation that accumu-
lations of these cells developed in association with
early emphysematous changes (21). More recently,
endotracheal instillations of homogenates of alveolar
macrophages have been reported to produce emphysema in
experimental animals (22). Although previous studies
failed to show significant levels of elastase in
alveolar macrophage lysosomes in comparison to those of
leukocytes (23,24), newer studies have demonstrated
that cultured alveolar macrophages secrete an elastase
into the medium, when stimulated by phagocytic challenge
or when harvested from the lungs of cigarette smokers
(18). If these observations are correct, the strategic
localization of most lung macrophages in the alveolar
air-spaces, coupled with their continual phagocytic
activity in that environment, renders them prime
candidates for a pathogenetic role in pulmonary emphysema.
   Neutrophilic leukocytes, on the other hand, also
gain access to lung connective tissues under both

normal and abnormal conditions, and these cells possess a very large amount of preformed elastase in their lysosomal (azurophilic) granules (9,10,25). Under physiologic conditions, neutrophils are sequestered in lung capillaries in large numbers and many of these cells emigrate from the blood and enter lung connective tissue interstitium. With shock, endotoxinemia or lung inflammation, the number of emigrating neutrophils is greatly increased. A pathogenetic role for neutrophil elastase in emphysema is suggested by the positive correlation between the amount of elastase in neutrophils of certain patients and the severity of their disease (26-28). In addition, neutrophil homogenates induce emphysema in experimental animals (22). More recently, a purified preparation of the human neutrophil enzyme has been shown to duplicate the effect of leukocyte homogenates (29,30).

Of the two cell types implicated in the pathogenesis of emphysema (alveolar macrophages and neutrophilic leukocytes), our laboratory has worked mainly on the latter. The principal theme of the following discussion, then, will be a review of our previous and current work on human neutrophil elastase in the context of the protease-pathogenesis model of emphysema.

II. PURIFICATION AND CHARACTERIZATION OF HUMAN NEUTROPHIL ELASTASE

Several methods have been published for isolation and purification of human neutrophil elastase (10-17). Our most recent techniques have been described in detail in an earlier report (12). Briefly, leukocytes are harvested from normal peripheral blood by dextran sedimentation and hypotonic lysis of red cells. The leukocytes are then suspended in 0.34 M sucrose and disrupted by mechanical means (passage through fine wire-mesh screen under suction) or by addition of heparin, and the granule fraction is finally collected by differential centrifugation. Granular proteins are extracted and crudely fractionated by freezing and thawing the granule pellet in buffers of varying salt concentrations (12). The elastase is nearly completely solubilized in 0.1M NaCl, buffered with 0.01M phosphate to pH 7.5. Affinity chromatography of this extract on a Sepharose column containing covalently-linked phenyl-butylamine (PhBuN-Affi-Gel) yields 90-100% pure neutrophil elastase with 60-70% recovery of starting enzyme (12).

Figs. 1 and 2 represent two typical elastase purification runs with 0.1M salt extracts of leukocyte granules. In both cases, the bulk of granular proteins was eliminated in the run-through fractions. Either of two procedures was then followed in order to obtain purified elastase. In one procedure (Fig. 1), column washing with 0.5M NaCl (pH 7.5) was continued until elastase activity, monitored by hydrolysis of tert-butyloxy-carbonyl-L-alanine-p-nitrophenyl ester (Boc-Ala-ONp), began to appear in the column effluent. Washing with this buffer was then continued until elastase activity of effluent fractions began, in turn, to decrease. At that time, the column wash was changed to 1.0M NaCl + 0.01M phosphate buffer, pH 7.5 + 20% dimethylsulfoxide (Me_2SO) (Fig. 1, arrow). This wash eluted a sharp peak of additional elastase activity from the column, as shown in Fig. 1.

Another procedure for recovery of purified elastase is shown in Fig. 2. In this method, column washing with 0.5M NaCl (pH 7.5) was stopped as soon as the bulk of chymotrypsin-like activity had passed through the column, and a 1.0M NaCl (pH 7.5) wash without Me_2SO was begun (see arrow A in Fig. 2). This wash was continued until Bz-Tyr-OEt esterase activity of the column effluent was no longer detectable, but before a significant amount of elastase activity had been eluted from the column. Thereafter, a Me_2SO gradient in 1.0M NaCl (pH 7.5) was begun (arrow B in Fig. 2). Under these conditions, the bulk of elastase eluted with that portion of the gradient between 5 and 12% Me_2SO and was obtained completely free of chymotrypsin-like activity. All traces of the latter had been eluted in the preceding 1.0M NaCl wash.

The polyacrylamide disc gel electrophoretic pattern given by the purified elastase fraction is shown in Fig. 3. A comparison of gel 2A in this figure (purified elastase) and gels 1A and 2 in Fig. 4 shows the degree

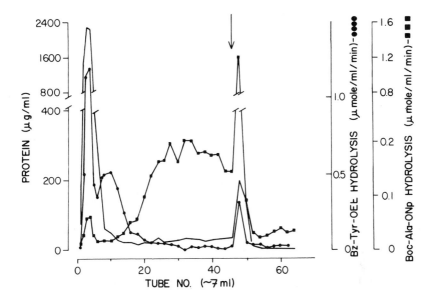

*Fig. 1.   Affinity chromatography of low salt (0.10
M NaCl) granular extract on PhBuN-Affi-Gel column.   The
PhBuN-Affi-Gel column (10 cm X 1.2 cm) was equilibrated
and run in the cold with pH 7.5, 0.01M NaPO$_4$ buffer
containing 0.5M NaCl.   The sample was applied to the
column in the same buffer.   Buffer change (arrow) was
to phosphate buffer containing 1.0M NaCl and 20% Me$_2$SO.
Flow rate, 20 ml/h.   Substrate: Bz-Tyr-OEt, ●——●;
Boc-Ala-ONp, ■——■.   (Bz-Tyr-OEt = Benzoyl-Tyrosine
Ethyl Ester) (Boc-Ala-ONp = t-butyloxycarbonyl-L-
alanine-p-nitrophenyl ester).   Reprinted from Feinstein,
G. and Janoff, A.   Bioch. Biophys. Acta, 403, 493
(1975).*

of isolation of the three well-characterized elastase
isoenzymes (see ref. 10-17) achieved by the affinity
chromatography procedures just described. The absence
of detectable chymotrypsin-like protease (cathepsin G)
in the purified elastase fraction was further substantiated by staining the gels with specific, chromogenic
ester substrates for the two proteases combined with
oxidized p-rosaniline (compare 1B and 1C in Figure 4
with 2B and 2C in Figure 3). The faint bands in gel
2C in Figure 3 in the elastase zone of the gel represent weak hydrolysis of the chymotrypsin substrate by
elastase.

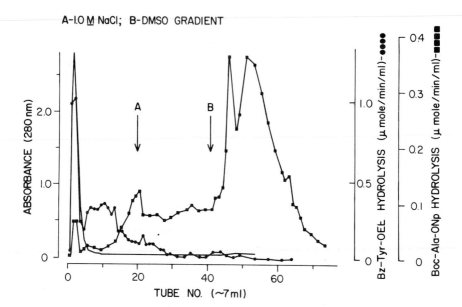

A-1.0 M NaCl; B-DMSO GRADIENT

Fig. 2. Affinity chromatography of low salt (0.10 M NaCl) granular extract on PhBuN-Affi-Gel column. The PhBuN-Affi-Gel column (10 cm X 1.2 cm) was equilibrated and run in the cold with pH 7.5, 0.01M NaPO₄ buffer containing 0.5M NaCl. The sample was applied to the column in the same buffer. First buffer change (arrow A): same NaPO₄ buffer containing 1.0M NaCl. Second buffer change (arrow B): 0-20% Me₂SO gradient in pH 7.5, 0.01M NaPO₄ containing 1.0M NaCl. Flow rate, 20 ml/h. Substrate hydrolysis: Boc-Ala-ONp, ▪; Bz-Tyr-OEt, ○──○. Reprinted from Feinstein, G. and Janoff, A. Bioch. Biophys. Acta, 403, 493 (1975).

Samples of purified enzyme were denatured by heating at 56°C for 90 min in the presence of 1% sodium dodecyl sulphate, 0.01M dithiothreitol and 0.001M EDTA and were then subjected to electrophoresis at pH 8.5 in 12% acrylamide gels containing 0.25% sodium dodecyl sulphate. The gels were calibrated with crystalline horseradish peroxidase (M_r=40,000) and crystalline soybean trypsin inhibitor (M_r=21,500), which had first been similarly denatured. An estimated M_r of 34,400

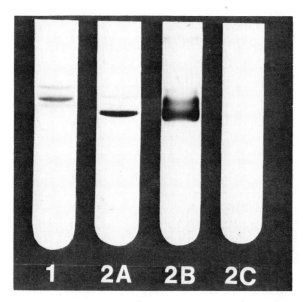

*Fig. 3. Acryalmide disc gel electrophoretograms.
Purified elastase fractions. Early elastase (1);
Elastase (2). Migration from top to bottom. 42.5 μg
of protein, Buffalo Black stain (1); 45 μg protein,
Buffalo Black stain (2A); 12.9 μg protein, Ac-DL-Ala-1-
ONap stain (2B); 43 μg protein, Ac-Phe-1-ONap stain
(2C). (Ac-DL-Ala-1-ONap = acetyl-alanine-naphthyl
ester, an elastase substrate; Ac-Phe-1-ONap = acetyl-
phenyl alanine-naphthyl ester, a chymotrypsin substrate)
Reprinted from Feinstein, G. and Janoff, A. Bioch.
Biophys. Acta, 403, 493 (1975).*

was calculated for the leukocyte elastase by this
method. This value agrees well with the M_r of 33,000-
36,000 estimated by Ohlsson and Olsson (11) for leukocyte elastase, but is larger than the values given by
Taylor and Crawford (13) or Baugh and Travis (14) of
22,000 and 26,000, respectively.

Table I presents the results of an amino acid
analysis of purified leukocyte elastase. On the basis
of the compositional analysis, a M_r of 34,970 was calculated which corresponds closely to the value obtained
from sodium dodecyl sulphate gel electrophoresis given
above. The amino acid analysis presented in Table I,

which is based on enzyme obtained from normal leukocytes, is compared with data previously published (11)
for human leukocyte elastase derived from leukemic
cells.

TABLE I   Amino Acid Analysis and Composition
of Human Granulocyte Elastase

Amino acid	nmol in analysis	mol/mol	% mol amino acids a*	b**
Lysine	10.3	3	1.0	0.5
Histidine	31.0	9	3.0	2.8
Arginine	92.5	27	9.0	11.7
Aspartic acid	99.6	29	9.7	9.6
Threonine	29.2	9	3.0	2.8
Serine	58.2	17	5.7	5.2
Glutamic acid	75.8	22	7.3	7.0
Proline	35.2	10	3.3	7.4
Glycine	99.0	29	9.7	11.2
Alanine	90.2	27	9.0	11.1
Half-cysteine	31.5	9	3.0	
Valine	74.4	22	7.3	11.4
Methionine	29.7	9	3.0	0.7
Isoleucine	61.5	18	6.0	4.2
Leucine	95.8	28	9.3	9.7
Tyrosine	43.8	13	4.3	1.1
Phenylalanine	66.2	19	6.3	3.9
Tryptophan	n.d.	n.d.	n.d.	n.d.
Total		300	100	100

*The amino acid composition we report herein was calcu-
lated assuming 3 mol of lysine residues per mol protein.
The results are also expressed as mol % for comparison
with previously published data by Ohlsson and Olsson
(11).*

 * = this study.
** = data from Ohlsson and Olsson (11), numbers refer
      to "Elastase I".
Reprinted from Feinstein, G. and Janoff, A.   Bioch.
Biophys. Acta, 403, 493 (1975).

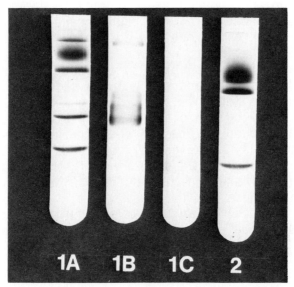

*Fig. 4. Acrylamide disc gel electrophoretograms.
Low salt (0.10M NaCl) crude granular extract (1) and
"runthrough" fraction from PhBuN-Affi-Gel chromatography (2). Migration from top to bottom. 1A and 1,
174 μg protein each, Buffalo Black stain; 1B, 116 μg
protein, Ac-DL-Ala-1-ONap stain; 1C, 116 μg protein,
Ac-Phe-1-ONap stain. Reprinted from Feinstein, G. and
Janoff, A. Bioch. Biophys. Acta, 403, 493 (1975).*

The purified human leukocyte enzyme hydrolyzed
several typical porcine pancreatic elastase substrates
such as Boc-Ala-ONp and acetyl-L-alanyl-L-alanyl-L-
alanine-ϕ-nitroanilide (Ac-(ala)$_3$-Nan). However, while
the pancreatic enzyme hydrolyzed Boc-Ala-ONp at twice
the rate that it cleaved Ac-(Ala)$_3$-Nan, this ratio was
20-fold higher in the case of the leukocyte elastase.
Fig. 6 shows the results of double immunodiffusion
tests employing purified leukocyte elastase and other
granule components vs. rabbit antisera raised against
the purified enzyme. The unispecific character of the
antiserum raised to purified leukocyte elastase is
evident from the fact that this antiserum gave single
lines of identity against pure elastase and crude
granular extract, but failed to react against purified
leukocyte chymotrypsin-like enzyme purified leukocyte

lysozyme or the PhBuN-Affi-Gel fraction containing the
run-through proteins of the crude granule extract.
(See Fig. legend for protein concentrations used.) In
a separate experiment, not shown here, crude granular
extract was first subjected to electrophoresis in an
acrylamide disc gel and the gel was sliced uniformly
throughout its length. Individual gel slices were then
set into wells cut into an agarose plate and double
diffusion allowed to proceed against the anti-elastase
antiserum. Only those gel slices containing the
elastase isoenzymes (as determined from a parallel
acrylamide gel stained with Buffalo Black) gave a
precipitin line.
        The unispecific antiserum and purified anti-
elastase antibodies isolated from it by absorption with
the pure enzyme were subsequently used for immunohistochemical localization of neutrophil elastase in lung
tissues (see sections V and VI, below). Non-immune
rabbit serum and absorbed antiserum were used as
controls.

III.   IN VITRO DEGRADATION OF LUNG ELASTIN BY PURIFIED
       NEUTROPHIL ELASTASE

        The solubilization of dog lung elastin in vitro
by purified neutrophil elastase is shown in Figure 7.
In this experiment, ten mg of dog lung elastin were
suspended in 1.0 ml of 0.01M sodium phosphate buffer,
pH 7.5, containing 0.25M NaCl, and an appropriate
amount of porcine pancreatic elastase or purified human
neutrophil elastase was then added. The mixture was
incubated with periodic shaking for 90 min at 37°C (the
same time interval that was subsequently used in whole
dog lung experiments). The reaction was terminated by
addition of an equal volume of cold 10 percent trichloroacetic acid and the acid-soluble peptides in the
trichloroacetic acid supernatant were measured by the
method of Lowry and co-workers (31), using a standard
curve of bovine serum albumin. On an equal weight
basis, the leukocyte enzyme was about 60 percent as
active as the pancreatic enzyme on the lung substrate.
On a molar weight basis, the neutrophil enzyme was

calculated to have 75 percent of the activity of the
pancreatic enzyme on this substrate.

ACRYLAMIDE GEL SCANS - BB STAIN

a - 0.10 M NaCl GRANULAR EXTRACT
b - PURIFIED GRANULAR ELASTASE

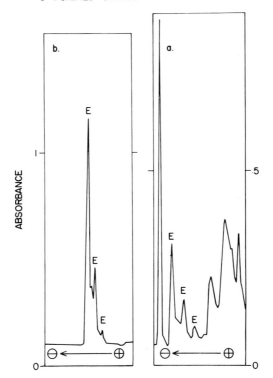

*Fig. 5. Scans of acrylamide disc gel electrophor-
etograms. a, low salt (0.10M NaCl) crude granular
extract, 130 µg protein; b, elastase, 45 µg protein.
Both stained with Buffalo Black stain. Reprinted from
Feinstein, G. and Janoff, A. Bioch. Biophys. Acta, 403,
493 (1975).*

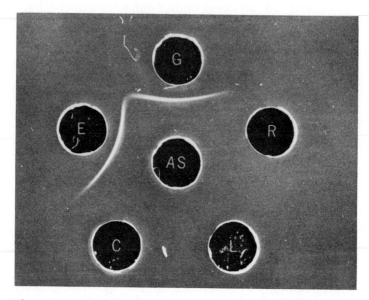

Fig. 6.   Double immunodiffusion between rabbit antiserum to leukocyte elastase and different components of the granular extract.   AS = anti-elastase antiserum, euglobulin-enriched fractions, 10 μl ; E = purified leukocyte elastase, 2.8 μg; G = crude granular extract (0.1M NaCl), 18.7 μg; R = run-through proteins of crude granular extract (PhBuN-Affi-Gel fractions), 17.4 μg; L = purified leukocyte lysozyme, 2.0 μg; C = purified leukocyte chymotrypsin-like enzyme, 3.9 μg. Reprinted from Feinstein, G. and Janoff, A.   Bioch. Biophys. Acta, 403, 493 (1975).

IV.   INDUCTION OF EXPERIMENTAL EMPHYSEMA IN THE DOG BY
      ENDOTRACHEAL INSTILLATION OF PURIFIED NEUTROPHIL
      ELASTASE

Having demonstrated that purified human PMN elastase was capable of producing significant <u>in vitro</u> digestion of dog lung elastin isolated from alveolar-enriched tissue, we next tested the enzyme's ability to induce emphysema-like lesions in whole lung. The first experiment was carried out in an isolated, perfused

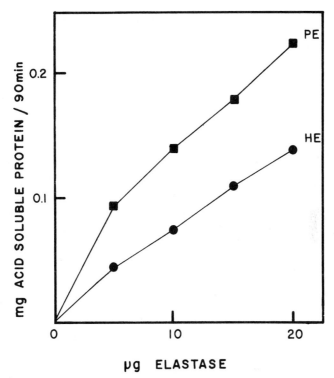

Fig. 7. *Digestion of dog lung elastin by pancreatic (PE) and neutrophil (HE) elastases, in vitro.
Milligrams of acid-soluble peptides released are plot-
ted on the ordinate versus micrograms of enzyme used on
the abscissa. All values have been corrected for zero
time and substrate minus enzyme controls. Reprinted
from Janoff, A. et al. Amer. Rev. Resp. Dis. 115, 461
(1977).*

lung ex situ. Pure bred beagles were killed by intravenous injection of sodium pentobarbital or intracard-
iac injection of potassium chloride and isolated lung
lobes were rapidly excised. After vascular perfusion
with saline, the lobes were lavaged with saline to
remove alveolar macrophages and extracellular anti-
proteinases. A No. 5 flow-directed, double-lumen Swan
Ganz catheter (0.8-ml capacity) was inserted into the
lobar bronchus and threaded into the smallest accepting

airway. The cuff was inflated and 1 ml of buffer
(control) or enzyme solution was gently delivered
through the inner lumen, followed by a 1-ml bolus of
air. After 5 min the cuff was deflated, the catheter
removed, the lobe inflated with air to a pressure of
25 cm H_2O, and the lobe incubated in the inflated state
at 22°C for a total of 90 min. After incubation, the
lobe was deflated and immediately reinflated with cold,
phosphate-buffered 10 percent formalin at a pressure of
25 cm H_2O for fixation before sampling tissues for
immunologic staining and histologic preparation. The
microscopic appearance of the lesions produced by two
different doses of PMN elastase 90 min after instillation of enzyme into isolated, perfused dog lung, as
described, is shown in Figure 8. Control tissue
exposed to buffer-vehicle alone and tissue exposed to
100 µg of pancreatic elastase are included in the
figure for comparison. It can be seen that, at a dose
of 384 µg, the PMN protease caused significant dilata-
tion of terminal respiratory structures, comparable to
that caused by 100 µg of pancreatic elastase in the
same lung.

Next, the induction of experimental emphysema by
purified human neutrophil elastase was attempted in
vivo. A purebred beagle was anesthetized with 0.5 ml
of acepromazine maleate (10 mg per ml) and 2.5 ml of
sodium pentobarbital (65 mg per ml), delivered intra-
venously. After insertion of an endotracheal tube, 3
Swan-Ganz catheters were placed in 3 different lobes
under fluoroscopy, and enzyme or buffer was instilled
into appropriate lobes. In the in vivo experiment,
the catheters were kept in place, cuffs inflated,
during the entire 90-min incubation. At the end of
this interval, the animal was killed by sodium pentobarbital, the chest quickly opened, and the lungs
perfused with chilled saline in situ through an opening
made in the right heart. The lungs were then removed
en bloc and inflated with cold 10 percent buffered
formalin to a pressure of 25 cm H_2O and fixed for 4
hours in the cold. Tissue samples were then cored from
the partially fixed lung to be used for immunologic and
ultrastructural studies. The remaining lung material

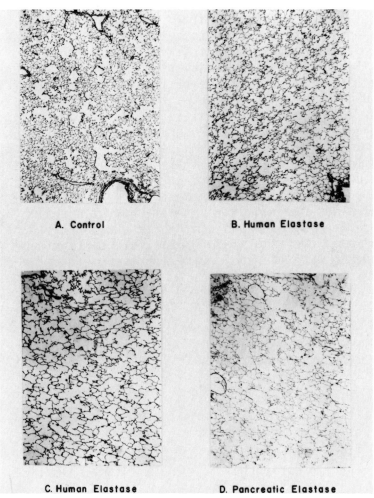

Fig. 8. <u>*In vitro*</u> *instillation. Histologic sections (6-μm thickness) of isolated, perfused dog lung instilled with various enzyme solutions. Paraffin-embedded sections stained with hematoxylin and eosin; original magnification: X 35. A. From a control tissue site treated with buffer-vehicle (1 ml). B. From a tissue site that received 115 μg of purified human polymorphonuclear leukocyte (PMN) elastase in 1 ml of buffer. C. From a tissue site that received 384 μg of the PMN enzyme. D. From a site that received 100 μg of pancreatic elastase in 1 ml of buffer. Reprinted from Janoff, A. <u>et al</u>. Amer. Rev. Resp. Dis. 115, 461 (1977).*

was fixed for an additional 20 hours and used for histologic analysis of the tissue immediately adjacent to
the cored areas.

Two different preparations of leukocyte elastase
were administered. In one of these, the enzyme purity
was 90 percent, its concentration was 1 mg per ml, and
a total of 1.33 mg were instilled. A separate lung
site was treated with a 97 percent pure preparation at
the same concentration (1 mg per ml), and a total of
2.52 mg of enzyme were instilled. Foci of alveolar
destruction were observable at both sites, 90 min after
enzyme instillation. Morphometric analysis of the
lesions was carried out by the method of Dunnill (32)
as modified by Johanson and Pierce (33). The quantitative measurements of mean linear intercept were in
agreement with the subjective impression gained from
the histologic examination, namely, that PMN elastase
had caused septal destruction (Table II).

TABLE II   Morphometric Analysis of In Vivo Experiment

Enzyme Dose (mg)	Tissue Zone*	Mean Linear Intercept Value	
		Normal Region†	Diseased Region†
1.33	Central	0.072	0.143
	Peripheral	0.075	0.116
2.52	Central	0.073	0.185
	Peripheral	0.068	0.130

*Tissue cores were biopsied from a central zone adjac-
ent to the tip of the instilling catheter and from a
peripheral site more distant from the catheter tip.

†Normal and diseased regions were selected according to
"patch analysis" method of Johanson and Pierce (33).

Reprinted from Janoff, A. et al. Amer. Rev. Resp. Dis.
115, 461 (1977).

The effects of the instilled enzyme on alveolar
septal ultrastructure in the dog are shown in Figs. 9
and 10. These figures are transmission electron micrographs of dog lung 90 min after buffer treatment
(control) or treatment with human polymorphonuclear
leukocyte (PMN) elastase, in vivo.

V. LOCALIZATION OF THE INSTILLED ELASTASE IN ENZYME-
TREATED DOG LUNG

Portions of lung were selected for enzyme localization studies after 4-hour fixation in cold phosphate-
buffered 10 percent formalin under a pressure of 25 cm
H_2O. Tissue was dissected from the partly fixed lung
and placed in 30 percent aqueous sucrose solution at
4°C overnight. Tissues were then quick-frozen by
immersion for 30 sec in isopentane cooled to liquid
nitrogen temperature, and were stored at -80°C in
precooled glass vials containing dry ice to prevent
dessication.
For electron microscopy, 40 m-thick frozen sections were cut and stained with the immunologic reagents as floating sections. Incubations were carried
out at 37°C, to facilitate penetration of reagents.
Purified, rabbit anti-PMN elastase antibodies (see
Section II, above) were diluted 1:20 in PBS. Sheep
antirabbit immunoglobulin antiserum (F(ab) fragments
conjugated with horseradish peroxidase) was diluted
1:40. Peroxidase reaction product was developed by
incubating the antisera-treated tissue slices for 7 to
10 min with a diaminobenzidine solution made up of 2 mg
of 3,3'-diaminobenzidine and 0.04 percent hydrogen
peroxide in 6.0 ml of 0.04M Tris/HCl buffer, pH 7.5.
Tissues were postfixed with glutaraldehyde and osmium
tetroxide and dehydrated. After dehydration, sections
were embedded in Epon and thin sections were cut by
ultramicrotome. Sections were stained with uranyl
acetate and lead citrate and examined using a JEM-7
transmission electron microscope operating at 80 kv.
Controls were tissue sections exposed to PMN elastase, which were treated with nonimmune rabbit serum or
absorbed antiserum as the primary reagent. Other

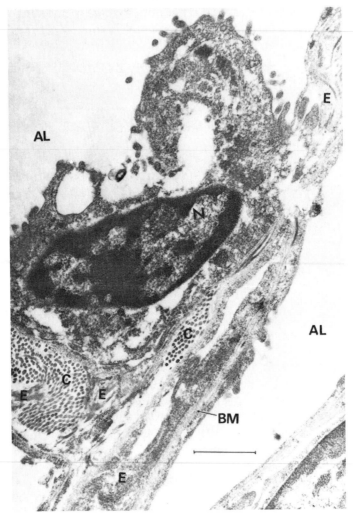

Fig. 9. Buffer-treated lung. Alveolar septum showing normal interstitial components including collagen bundles (C), elastin (E) and basement membrane (BM). AL = alveolar airspace. N = epithelial cell nucleus. Note intact septal structure. Uranyl acetate and lead citrate counterstain. A 1-μm marker is included for reference. Reprinted from Janoff, A. et al. Amer. Rev. Resp. Dis. 115, 461 (1977).

controls were tissue sections exposed to the buffer-
vehicle alone and then treated with anti-elastase antibodies as the primary reagent. Control tissues were
uniformly negative. Background activities of lung
endogenous peroxidase and pseudoperoxidase (erythrocytes) were abolished by preincubation of tissues in
absolute methanol for 20 min, followed by a rinse in
0.125 percent hydrogen peroxide in PBS for 20 min.

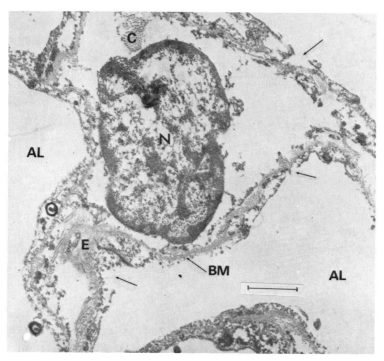

*Fig. 10. Enzyme-treated lung. Alveolar septum
showing loss of interstitial components, except for
residual collagen bundles (C) and a small amount of
elastin (E). Numerous discontinuities in septal wall
(arrows) are apparent. Note disrupted septal structure.
See Fig. 9 for definitions of other abbreviations.
Reprinted from Janoff, A. et al. Amer. Rev. Resp. Dis.
115, 461 (1967).*

Figures 11 and 12 show the results of the immunoperoxidase staining to localize the instilled leukocyte
enzyme in dog lung. Attachment of the elastase to lung
elastic fibers can be seen in both Figures.

VI. IDENTIFICATION OF NEUTROPHIL ELASTASE IN HUMAN
LUNG TISSUE

In human lung, neutrophil elastase may be liberated endogenously from marginated PMN in lung capillaries, from diapedesing leukocytes, from emigrated cells,
or from leukocytes that have reached the air-spaces and
are engaged in phagocytosis. Lung PMN have been
observed to undergo degranulation under a variety of
conditions (34), an event which normally accompanies
release of lysosomal enzymes. Using the same techniques that were previously employed to visualize exo-
genously instilled neutrophil elastase in dog lung, we
set out to detect endogenously released neutrophil
elastase in human lung. Some of the preliminary
results of these indirect immunoperoxidase stains are
presented next.

Tissue samples were obtained from four separate
lungs and cooled to 4°C immediately after resection.
Specimens were trimmed and quick-frozen in isopentane
cooled to liquid nitrogen temperature, within 1-2 hours
of surgery. They were thereafter stored at -80°C in
tightly-stoppered, pre-chilled vials containing solid
CO_2 to prevent dessication. Frozen sections were cut
at 6 μm thickness on a cryostat. Immunohistochemical
procedures were then carried out at the level of light
microscopy. Endogenous peroxidase activity of lung
cells and of neutrophils was suppressed as before (see
Section V), and the same control stains were included
as in the dog lung experiments. These proved uniformly
negative.

Fig. 13 shows selected results. In the upper left
illustration (original magnification = 400X), a number
of PMN are visible within the lung specimen. Nuclear
lobes are unstained, while cytoplasm is strongly positive for elastase (myeloperoxidase was suppressed).
Examination of individual leukocytes at higher magnification (X1000) reveals apparent extracellular leakage

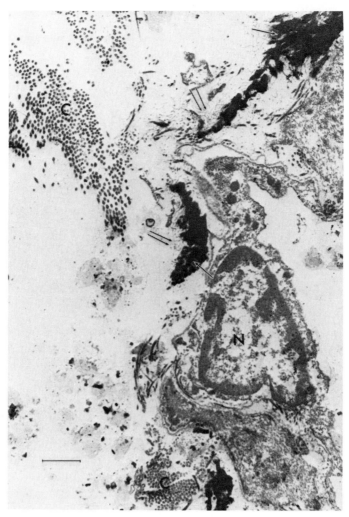

*Fig. 11. Electron micrograph showing peroxidase
reaction product distributed throughout lung connective
tissue. Enzyme-treated lung. Rabbit antibody to neutrophil elastase as primary stain. Note microfibrillar
structures (small double arrows) surrounding dense
deposits of reaction product (small single arrow),
which appears to be covering the amorphous component of
elastic fibers. Collagen fibers (C) are relatively
free of peroxidase reaction product. N = epithelial
cell nucleus. Reprinted from Janoff, A. et al. Amer.
Rev. Resp. Dis. 115, 461 (1977).*

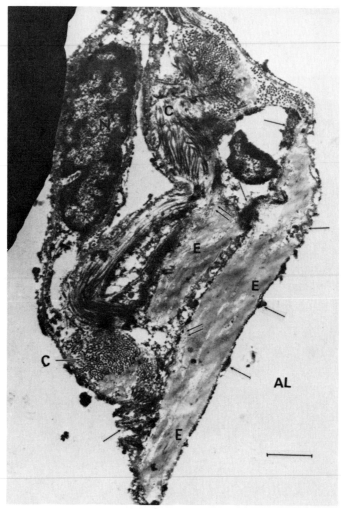

Fig. 12. Enzyme-treated lung. Anti-elastase antibody (rabbit) was used as primary stain. Two elastic
fibers (E) are visible with attached peroxidase reaction
product present at many locations (single small arrows),
signifying adherent enzyme at these sites. The amorphous elastin component of the elastic fibers appears
attenuated in several regions, exposing structures that
may represent microfibrillar components of the elastic
fiber (double small arrows). Intact collagen bundles
(C) and a cell nucleus (N) are also visible. Reprinted
from Janoff, A. et al. Amer. Rev. Resp. Dis. 115, 461
(1977).

of elastase from some of the cells (Fig. 13, upper
right). However, the relatively poor preservation of
structural detail in these uninflated, atelectatic
frozen sections makes precise localization of the cells
impossible. At other sites, leukocyte elastase can be
identified in the connective tissues in clearly extracellular patterns (Fig. 13, lower left, arrow). Such
patterns were found in one lung specimen obtained from
a patient with advanced emphysema (bullous resection).
No conclusions can be drawn regarding the activity of
the extracellular elastase, however, since elastase-
alpha$_1$ proteinase inhibitor complexes also react with
anti-elastase antibodies (35). Indeed, other sections
of lung tissue from this same specimen also stained
positively with an antiserum to human alpha$_1$-proteinase
inhibitor (Fig. 13, lower right). Non-immune rabbit
serum controls were consistently negative for all lung
sections.

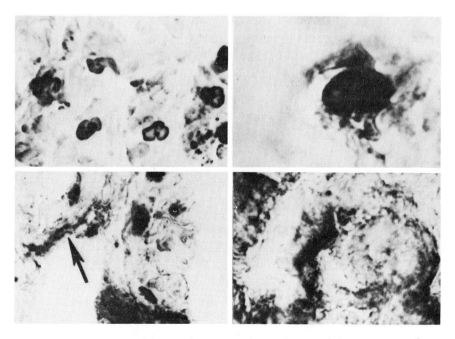

*Fig. 13. Light micrographs of freshly resected
human lung tissues stained for neutrophil elastase and
alpha$_1$ proteinase inhibitor by the indirect immunoperoxidase technique.*

These observations suggest that PMN elastase may be released in human lung during life, and contribute to the destruction of alveolar walls seen in emphysema. Alternatively, leakage of enzyme in the specimens we tested could also have resulted from PMN-injury caused by anesthesia or trauma associated with surgical removal of the lung. In addition, some PMN autolysis may have occurred in spite of rapid refrigeration of the freshly resected tissue. For such reasons, caution should be exercised in the interpretation of these highly preliminary findings.

VII.   POSSIBLE MECHANISMS OF EMPHYSEMA IN CIGARETTE SMOKERS, VIEWED IN THE CONTEXT OF THE PROTEASE-PATHOGENESIS HYPOTHESIS AND THE ROLE OF NEUTROPHIL ELASTASE

Cigarette smoking is recognized to be a major risk factor associated with the development of pulmonary emphysema in man (4). It would therefore seem appropriate to search for mechanisms by which cigarette smoking might favor excess of free proteases in the lung. Table III outlines some of the mechanisms that can be suggested.

We recently carried out studies (37) designed to test possibility (C), as listed in Table III. To investigate suppression of protease inhibition by cigarette smoke, elastin-agarose gels were impregnated with cigarette smoke condensate dissolved either in dimethyl-sulfoxide or in aqueous solutions of pulmonary surfactant. Control gels were impregnated with the solvents alone. Smoke condensate affected neither local pH in the gel nor diffusion of subsequent reactants. Elastases from various sources were then allowed to diffuse through the impregnated gels towards a counter-diffusing sample of anti-protease. The effectiveness of the anti-protease in blocking the enzyme was determined from the extent of elastolysis. The elastin substrates used included beef ligament elastin and dog lung elastin. The enzymes employed were porcine pancreatic elastase and purified human leukocyte elastase. The anti-proteases tested included whole human serum, purified

TABLE III   Protease-Pathogenesis Model of Emphysema

Possible Relationships to Smoking

Smoke components may--

(A)   Increase Lung Proteases
        PMN recruitment (see ref. 36)
        Stimulate PMN enzyme release*
        Stimulate PAM enzyme secretion (see ref. 18)
        Decrease lung barriers to circulating proteases

(B)   Increase Digestibility of Lung Target Structural
      Macromolecules
        Elastin
        Collagen
        Proteoglycan

(C)   Decrease Effectiveness of Lung Antiproteases
        Alpha$_1$ Proteinase Inhibitor
        Alpha$_2$ Macroglobulin
        Alpha$_1$ Antichymotrypsin
        Bronchial mucous inhibitor
        Other endogenous lung inhibitors

(D)   Decrease Containment or Clearance of Protease-
      Antiprotease Complexes from Lung--
        by Mucociliary stream
        by Macrophages

(E)   Decrease Lung Connective Tissue Repair

(F)   Other?

---

*Preliminary evidence from our laboratory supports this
possibility (M. Blue and A. Janoff, "Possible Mechanisms of Emphysema in Smokers. II. Release of Elastase
from Human Neutrophilic Leukocytes by Cigarette Smoke
Condensate In Vitro and In Vivo," Ms. in preparation).

human alpha$_1$-proteinase inhibitor, crude human bronchopulmonary lavage fluid, and a synthetic chloromethyl
ketone inactivator of elastase. Whole, unfractionated
cigarette smoke condensate suppressed elastase inhibition by the endogenous anti-proteases, but failed to
affect the chloromethyl ketone. Nicotine and acrolein,
tested alone, did not duplicate the suppressive effect
of whole smoke condensate. The suppressive action of
cigarette smoke on whole human serum was prevented by
the reducing agent, glutathione, suggesting that
oxidants or other electrophilic agents in cigarette
smoke might be responsible for the observed effect.
Indeed, others have shown that oxidation of methionine
residues in chicken ovoinhibitor completely blocks the
elastase-inhibitory activity of that agent (38). Some
of the foregoing results are depicted in Figs. 14-17.

VIII. CONCLUDING REMARKS

The preceding sections briefly reviewed our work
on the purification and characterization of human leukocyte elastase and on its ability to attack lung elastin
in vitro and in vivo. In the latter case, we were able
to show that the enzyme became bound to lung elastic
fibers and caused destruction of alveolar walls. Our
suggestion that similar mechanisms might contribute to
alveolar injury in pulmonary emphysema was supported by
two additional observations, also summarized in the
present report. The first was the immunological demonstration of leukocyte elastase in connective tissue
interstitium of freshly excised human lung. The second
was the demonstration that cigarette smoke can suppress
those endogenous inhibitors of leukocyte elastase which
are important in the defense of the lung against proteo-
lytically-mediated damage.
It must be emphasized that our findings do not
exclude a complementary role for other neutrophil hydrolases (including other neutral proteases) in the pathogenesis of the lung injury associated with emphysema;
nor have we excluded a supplementary role for elastolytic proteases derived from other cell sources in this
disease. Chief among the latter would be the alveolar

Fig. 14. Effect of cigarette smoke condensate (CSC) on inhibition of porcine pancreatic elastase by purified human alpha₁-proteinase inhibitor. A = 1.2 µg alpha₁-proteinase inhibitor, C = 1 mg CSC in DMSO, D = DMSO alone, E = 0.4 µg pancreatic elastase. Wells labelled C and D were filled 4 hr before the enzyme and inhibitor wells. Substrate = bovine ligament elastin conjugated with fluorescein (elastin particle size below 37 µm). The agarose gel was buffered to pH 8.8 with 0.2M Tris-HCl, containing 0.2 percent sodium azide. Incubation time at 37°C = 16 hr. Reprinted from Janoff, A. and Carp, H. Amer. Rev. Resp. Dis.

macrophage, whose probable contribution to lung injury in emphysema was already reviewed in an earlier section of this paper. In addition, platelet elastase (19) and perhaps even circulating elastase of pancreatic origin may play a pathogenetic role.

It is also worth noting that neutrophil elastase has other important targets in the lung besides elastin, and that enzymatic attack upon these alternative targets could also explain part of the lung injury caused by the protease. For example, neutrophil elastase has been shown to degrade proteoglycans of joint cartilage (39, 40) suggesting that proteoglycans of airway cartilage and of lung parenchymal ground substance represent

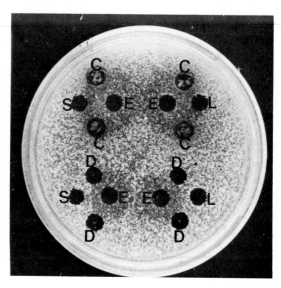

Fig. 15. Effect of CSC on inhibition of human neutrophil elastase by human serum and bronchopulmonary lavage fluid. S = 14 percent whole human serum, L = crude human lung lavage containing 0.4 percent protein, C = 2 mg CSC in DMSO, D = DMSO alone, E = 10 μg leukocyte elastase. C and D wells were preloaded as before. Substrate was dog lung elastin (particle size below 40 mesh). The gel was buffered to pH 8.6 with 0.4M Tris-HCl containing 0.001M Ca^{2+} and Mg^{2+}, 0.017M beta-alanine and 0.02 percent sodium azide. Incubation time = 72 hr. Reprinted from Janoff, A. and Carp, H. Amer. Rev. Resp. Dis.

other important targets for the enzyme. Vascular basement membrane appears to be another alternative connective tissue substrate for the neutrophil enzyme (41,42). Thus, lung basement membranes (or at least the endothelial component thereof) may also be injured. Recently, evidence has been obtained which shows that leukocyte elastase can also digest the apoproteins of pulmonary surfactant (King, R. and Janoff, A., unpublished observations). Thus, the neutrophil enzyme may also damage soluble protein constituents of the lung, in addition to supporting fibers and membranes, and may

cause important functional as well as structural de-
rangements in that organ.

For all these reasons, further study of neutrophil
elastase as a mediator of tissue damage in lung diseases
seems worthwhile. In particular, its role in the
protease-pathogenesis model of pulmonary emphysema
deserves continued exploration.

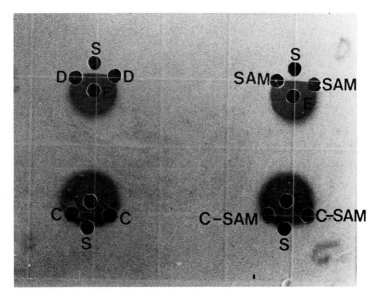

*Fig. 16. Effect of CSC dissolved in aqueous
suspension of pulmonary surfactant on inhibition of
pancreatic elastase by whole human serum. S = 10
percent human serum, E = 0.4 µg pancreatic elastase,
C = 1 mg CSC dissolved in DMSO, D = DMSO alone, C-SAM =
CSC sonicated in 10 percent surfactant suspended in
water, SAM = 10 percent surfactant alone. (The concentration of CSC in the aqueous surfactant solution was
estimated to be one-fourth of that present in the DMSO
solution, based on absorbance at 410 nm). All other
conditions as in Fig. 13.*

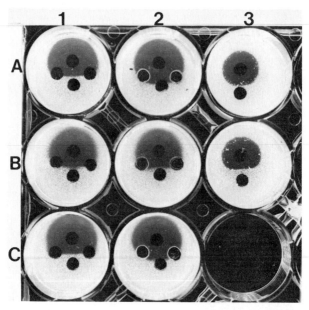

Fig. 17. Reversal of CSC-suppressor effect by a reducing agent, and duplication of CSC-suppressor effect by an oxidant. Description of plates: C-1, standard inhibition of elastase by serum, when DMSO is preloaded in plate. C-2, standard suppression of serum-inhibition, when CSC is preloaded in plate. B-1, same as C-1 except 50 mM GSH present. B-2, same as C-1 except 50 mM GSH present. A-1, same as C-1 except 5 mM GSH present. A-2, same as C-2 except 5 mM GSH present. Note that 50 mM GSH (reduced glutathione) prevents the suppression of serum-inhibition normally given by CSC (plate B-2); but at 1/10th this concentration (plate A-2), the protective effect of GSH is lost. A-3, elastase vs. serum in a plate equilibrated with chloramine T. B-3, elastase vs. serum without chloramine T. The last two plates were incubated for a shorter interval than the others. Note the inhibition of elastase by serum in B-3, and the suppression of serum-inhibition by the oxidant in A-3. Concentration of serum = 10 percent in all plates; amount of elastase (porcine pancreatic enzyme) = 0.4 µg in all plates.

IX. REFERENCES

1. "Pulmonary Emphysema and Proteolysis" (C. Mittman, Ed.), Academic Press, New York, 1972.
2. Kaplan, P.D., Kuhn, C., and Pierce, J.A., J. Lab. Clin. Med. 82, 349 (1973).
3. Eriksson, S., Acta Med. Scand. 177 (Suppl. 432), 1 (1965).
4. Auerbach, O., Hammond, E.C., Garfinkel, L., and Benante, C., New Eng. J. Med. 286, 853 (1972).
5. Blackwood, C.E., Hosannah, Y., Preman, E., Keller, S., and Mandl, I., Proc. Soc. Exp. Biol. Med. 144, 450 (1973).
6. Snider, G.L., Hayes, J.A., Franzblau, C., Kagan, H.M., Stone, P.S., and Korthy, A.L., Am. Rev. Resp. Dis. 110, 254 (1974).
7. Kuhn, C., Yu, S.Y., Chraplyvy, M., Linder, H.E., and Senior, R.M., Lab. Invest. 34, 372 (1976).
8. Kuhn, C., and Starcher, B., Am. Rev. Resp. Dis. 113 (Part 2), 209 (1976).
9. Janoff, A., and Scherer, J., J. Exp. Med. 128, 1137 (1968).
10. Janoff, A., Lab. Invest. 29, 458 (1973).
11. Ohllson, K., and Olsson, I., Eur. J. Biochem. 42, 519 (1974).
12. Feinstein, G., and·Janoff, A., Biochim. Biophys. Acta 403, 493 (1975).
13. Taylor, J.C., and Crawford, I.P., Arch. Biochem. Biophys. 169, 91 (1975).
14. Baugh, R.J., and Travis, J., Biochemistry 15, 836 (1976).
15. Rindler, R., Schmalzl, F., and Braunsteiner, H., Schweiz. Med. Wschr. 104, 132 (1974).
16. Gerber, A. Ch., Carson, J.H., and Hadorn, B., Biochim. Biophys. Acta 364, 103 (1974).
17. Schmidt, W., and Havemann, K., Hoppe-Seyler's Z. Physiol. Chem. 355, 1077 (1974).
18. White, R., Hsui-San, L., and Kuhn, C., Fed. Proc. 36, (3) 1262 (1977).
19. Robert, B., Legrand, Y., Pignaud, G., Gaen, J., and Robert, L., Path.-Biol. 17, 615 (1969).
20. Balo, J., and Banga, I., Nature (Lond.) 164, 491 (1949).

21. Heppleston, A.G., J. Path. Bact. 66, 235 (1953).
22. Mass, B., Ikeda, T., Meranze, D.R., Weinbaum, G., and Kimbel, P., Am. Rev. Resp. Dis. 100, 384 (1972).
23. Janoff, A., Rosenberg, R., and Galdston, M., Proc. Soc. Exp. Biol. Med. 136, 1054 (1971).
24. Levine, E.A., Senior, R.M., and Butler, J.V., Am. Rev. Resp. Dis. 133, 25 (1976).
25. Dewald, B., Rindler-Ludwig, R., Bretz, V., and Baggiolini, M., J. Exp. Med. 141, 709 (1975).
26. Galdston, M., Janoff, A., and Davis, A.L. Am. Rev. Resp. Dis. 107, 718 (1973).
27. Rodriguez, J.R., Seals, J.E., Radin, A., Lin, J.S., Mandl, I., and Turino, G.M. Clin. Res. 23, 349a (1975).
28. Crawford, I., in "Isolation, Characterization and Mechanism of Action of Proteases and Antipro-teases", Division of Lung Diseases Workshop (1976).
29. Janoff, A., Sloan, B., Weinbaum, G., Damiano, V., Sandhaus, R.A., Elias, J., and Kimbel, P. Am. Rev. Resp. Dis. 115, 461 (1977).
30. Senior, R., Am. Rev. Resp. Dis. In Press.
31. Lowry, O.H., Rosebrough, N.J., Farr, A.L., and Randall, R.J., J. Biol. Chem. 193, 265 (1951).
32. Deunnill, M.S., Thorax 17, 320 (1962).
33. Johanson, W.G., and Pierce, A.K., J. Clin. Invest. 51, 288 (1972).
34. Wilson, J.W., Ratliff, N.B., Mikat, E., Hackel, D.B., Young, W.G., and Graham, P.C., Chest 59, 36s (1971).
35. Ohlsson, K., in "Proteases and Biological Control" (E. Reich, D.B. Rifkin, and E. Shaw, Eds.), Cold Spring Harbor Press, 591 (1975).
36. Kilbern, K.H., and McKenzie, W., Science 189, 634 (1975).
37. Janoff, A., Carp, H., Am. Rev. Resp. Dis. In Press.
38. Schechter, Y., Burstein, Y., and Gertler, A., Biochem. 16, 992 (1977).
39. Janoff, A., Feinstein, G., Malemud, C.J., and Elias, J.M., J. Clin. Invest. 57, 615 (1976).

40.   Keiser, H., Greenwald, R.A., Feinstein, G., and
      Janoff, A., J. Clin. Invest. 57, 625 (1976).
41.   Janoff, A., and Zeligs, J.D., Science 161, 702
      (1968).
42.   Janoff, A., Lab. Invest. 22, 228 (1970).

DENTAL CARIES

William H. Bowen

Caries Prevention and Research Branch
National Caries Program
National Institute of Dental Research

Dental caries is probably the most prevalent disease in the United States, and costs the population several billion dollars annually. Caries results from the interaction of specific micro-organisms with dietary carbohydrate on the tooth surface. The formation of dental plaque is probably the first evidence of this interaction. Following each ingestion of sugar the pH value of plaque falls to values as low as 4.0 which results in rapid demineralization of enamel. Evidence indicates that dental caries is an infectious and transmissable disease in humans even though dietary substrate is essential in its pathogenesis. It appears that a number of mechanisms are involved in the colonization of the tooth surface by bacteria; both ionic and hydrophobic bonding are probably involved. Several approaches are available for the prevention of dental caries. Water fluoridation and the restriction of the frequency of intake of fermentable carbohydrate are the most effective. Vaccines, antimicrobial agents and novel methods of delivering them are being researched.

Dental caries is one of the more common afflictions of humans. In many countries more than 95% of the
population is affected, and the number of new lesions
is increasing more rapidly than they can be treated.
Each year the population of the United States spends
approximately 7 billion dollars to have the ravages of
dental decay repaired. This expenditure represents but
a small fraction of the overall need. Although the
expenditure for treatment is large, it does not
represent the total economic consequence of this
disease. It is difficult to estimate the number of
working days lost through dental disease, or the number
of sleepless nights resulting from pain and suffering
(1).

It is clear that dental caries is a major public
health problem and that considerable efforts need to be
expended to understand its pathogenesis and to bring it
under control. Repair of rotting teeth has become so
dominant a part of dental practice that tooth decay
has come to be regarded as inevitable. It requires a
conscious effort to realize that tooth decay is not
normal.

Dental caries results from the interaction of
specific microorganisms on susceptible tooth surfaces
with specific components of the diet, particularly
carbohydrate in the form of sugar. In the absence of
any one of these factors caries will not develop. The
formation of dental plaque is the first evidence of
this interaction.

Dental plaque is a soft, adherent accummulation of
microorganisms embedded in a matrix of carbohydrate and
protein (2). The biochemical composition of plaque is
receiving an increasing amount of attention because it
is apparent that the ability of plaque to produce
disease should be reflected in its biochemical composition (3).

Study of the composition of plaque is difficult,
because the levels of its various constitutents are
influenced by the duration it has been allowed to
accummulate, the site from which the samples are
removed, the composition of the diet, the time elapsed
since the last intake of food, and the microbial
composition (4,5).

Plaque in situ is composed of almost 80% water. This percentage aqueous composition appears to vary little even from site to site either in humans or primates. The majority of investigations have been confined to observations on whole plaque either wet or dried. Few determinations have been made on the fluid phase of plaque which may be a major omission because it appears likely that the pathogenicity of plaque would be expressed in the composition of plaque fluid (6,7).

The bulk of the solid phase of plaque is comprised of protein (35-47%) and specific proteins have been isolated from plaque and identified. Its levels of amylase, lysozyme, and albumin have been determined (8), and immunoglobulins G and A have also been measured (9). IgA represents approximately 1.6-2.7% of extractable protein.

In contrast to the protein content, the amount and type of carbohydrate found in plaque is influenced by the composition of the diet (10,11). The bulk of plaque polysaccharide is found extracellularly and contributes 10-20% to the volume of plaque (12,13,14). Microorganisms in plaque have the ability to synthesize glucan and fructan from sucrose using the enzyme glucosyltransferase and fructosyltransferase (15). In persons consuming a diet rich in sucrose, glucan is by far the predominant polysaccharide in plaque.

The predominant glucan in plaque was observed to have a structure of primarily α-1,3 linkages, and that only a small proportion of the glucan had a structure of α-1,6 linkages (15). The term mutan has been proposed for the material containing α-1,3 linkages because a similar substance is synthesized by Streptococcus mutans from sucrose.

Dental plaque contains 1-2% fructan; following ingestion of sucrose significantly higher levels may be present. Many plaque bacteria have the ability to metabolize fructans (16,17).

It is reasonable to assume that the levels of calcium and phosphorus in plaque play a crucial role in the pathogenesis of dental caries if only because enamel is composed largely of calcium and phosphorus. Plaque appears to have the capacity to concentrate calcium and phosphorus because levels found in it are in excess of those present in saliva (18). The state of calcium and phosphorus in plaque is unclear. The proportion of ionic calcium to bound calcium does not appear to have been determined. It is probable, however, that a considerable fraction is bound because many microorganisms in plaque have the capacity to calcify, and it has been shown (19) that this process starts intracellularly. The level of phosphorus in plaque is dependent on the site from which samples are removed; the proportion of inorganic P to total P is also site dependent (18).

Composition of diet also influences the calcium and phosphorus levels of plaque. Monkeys which received their entire diet by gastric intubation or received only casein orally had significantly more calcium and phosphorus in their plaque than animals which ingested sucrose or glucose and fructose orally (20). The calcium level in 'starved' plaque declined from 7.44 μg/mg dry weight to 5.76 following a 5 minute exposure to sugar. The inorganic phosphorus declined from 6.11 to 4.30 and the total phosphorus from 14.79 to 12.69 in the same investigation (18).

Plaque also apparently has the ability to concentrate fluoride although there is little agreement on the levels found there. Nevertheless, it is apparent that people who reside in a fluoridated area have substantially more fluoride in their plaques than those in a non-fluoridated area. The concentrations of fluoride in plaque vary from 8 ppm to as high as 130 ppm (21, 22). Plaque fluoride is not derived from enamel. Monkeys which received a daily dose of 2 mg F by gastric intubation had significant levels of fluoride in their plaque, indicating that the saliva and possibly gingival crevice fluid provide the main routes for fluoride to enter plaque.

Since the time of W.D. Miller (23) it has been generally accepted that microorganisms are essential in the pathogenesis of dental caries. The etiologic role of Streptococcus mutans was first suggested more than 50 years ago by J.K. Clarke (24) who wrote "the evidence suggests that caries is due to infection of the teeth in certain circumstances not yet made clear by a

hitherto undescribed steptococcus S. mutans." With few
exceptions, for example those of Fish (25) and McClean
(26), investigators paid little attention to Clarke's
observations for the next 43 years. Indeed, Clarke
apparently did not publish additional papers on this
topic.

Because of the apparent importance of acid production in the pathology of dental caries, considerable
attention was focused on the role of lactobacilli in
the etiology of dental caries (27). For almost three
decades the role of lactobacilli in the pathogenesis
of dental caries was investigated to the exclusion of
other microorganisms (28,29).

The resurgence of interest in S. mutans was
heralded by Keyes' observation that dental caries is an
infectious and transmissible disease, in hamsters at
least (30). This observation was elegantly demonstrated by carrying out the following investigation. One-
half of a litter was taken from a strain of hamsters
which had been caries-active for four generations, and
rendered caries-free by including antibiotics in its
diet, while the other half served as controls. A
litter was then bred from a female in each group. Half
of the offspring from the caries-inactive animal was
housed alone while the other half was housed with the
caries-active controls. The isolated animals remained
caries-free, while those placed with the controls
developed dental caries. Shortly after this investigation, Fitzgerald and Keyes (31), Gibbons (32) and
Krasse (33) showed that rodents infected by specific
organisms subsequently identified as S. mutans
developed dental caries. Zinner (34), Krasse (33) and
Gibbons (32) and their colleagues demonstrated that
hamsters and rats infected with S. mutans of human
origin and fed a cariogenic diet developed rampant
dental caries. These observations were confirmed and
extended by Bowen who observed that young monkeys
infected with S. mutans and fed a diet, all the components of which are normally purchased for human consumption, developed caries which clinically, radiographically and histologically could not be distinguished
from dental caries occurring in humans (35).

These observations could be particularly relevant
to the situation which prevails in humans. Epidemio-
logical evidence gathered by Klein (36), Böök and
Grahnen (37), and Davies (38) shows that dental caries
patterns of children tend to follow that of their
parents and particularly those of their mother. It is
true, of course, that a non-bacteriological explanation
for this phenomenon can be offered. However, Berkowitz
and colleagues (39) have shown that S. mutans found in
the mouths of infants appears to derive primarily from
their mothers.

S. mutans possesses several interesting characteristics which probably equip it to be cariogenic. S.
mutans is rarely found in the mouths of newborn humans
or primates. Carlson observed that S. mutans dis-
appears from the mouths of edentulous patients following removal of their dentures, and reappears when
dentures are reinserted (40). S. mutans is not found
in either the mouths or feces of young children until
teeth erupt, and the numbers isolated increase as more
teeth appear (H.M. Stiles, personal communication); a
similar observation was made by Cornick and Bowen (41)
in monkeys. It is clear, therefore, that S. mutans
requires a solid surface to become established in the
mouth, and should probably be classified as an obligate
periphyte.

The mechanism by which S. mutans becomes attached
to the tooth surface has been the subject of intensive
research over recent years. It is clear that S mutans
can colonize tooth surfaces in the absence of sucrose
even though there appeared evidence which suggested the
contrary. Subjects on a sucrose-free diet harbor S.
mutans (42), and Guggenheim, Bowen and Thomson
(unpublished work) have also shown the presence of
S. mutans in the plaque of fructose intolerant patients
who must abstain from sucrose use.

There is little direct evidence that bacteria
colonize the tooth surface directly, although there is
no reason why they should be incapable of doing so. It
is probable that oral microorganisms colonize salivary
pellicle which is rapidly formed on cleaned tooth
surfaces. It has been shown that pellicle is rich in
acidic groups (43). Krogstad et al. (44) have shown

that highly acidic sulphated glycoproteins may be specifically adsorbed onto the tooth surface and that the
sublingual glands appear to be their primary source.
This adsorption results in a net negative charge on the
tooth surface which should, theoretically at least,
make it difficult for the negatively charged microorganisms to colonize the tooth surface. It has been postulated that calcium bridges are formed between the negative charges resulting in colonization. The nature of
the negative charge on the surface of S. mutans has
been the subject of some investigation. Evidence
indicates that phosphate is, in part at least, responsible and that it is probably present in the form of
lipoteichoic acid (45). S. mutans grown in the presence of sucrose carries a much greater negative charge
than the microorganisms grown in the presence of glucose, an observation which may explain the ability of
sucrose to enhance colonization of tooth surfaces by
S. mutans (46). Because of the apparent predilection
of S. mutans to colonize solid surfaces, it appears
unlikely that ionic interactions alone explain the
phenomenon. Saliva, in addition to enhancing the
receptivity of the enzyme surface to colonization, can
influence the adherence of microorganisms in other
ways. Many investigators have noted that oral microorganisms aggregate in the presence of saliva. It
also appears that these aggregating factors are species
specific. Adsorption of saliva by one type of organism
does not necessarily remove the ability of the saliva
to aggregate other microorganisms. Available evidence
indicates that sIgA is not involved in this phenomenon.
These agglutinating factors appear related to the proteins which are adsorbed onto tooth hydroxyapatite and
are termed salivary lectins (48). It also appears
that many plaque bacteria can agglutinate red blood
cells, an observation which may explain in part the
affinity of some oral bacteria for pellicle. Pellicle
has been shown to contain blood group substances (49).
It appears possible that hydrophobic bonding, too, may
be implicated (50).

     S. mutans and other microorganisms found in dental
plaque have the ability to synthesize extracellular
polysaccharide glucan and fructan from sucrose and

possibly also from other sugars. Both water soluble and insoluble glucan are formed. Results of structural analyses reveal that the water soluble glucan is primarily α-1,6 linked glucose; in contrast, the water insoluble glucan is essentially α-1,3 linked glucose (15). The ability to form water insoluble glucan appears to be intimately related to the pathogenic capacity of S. mutans. deStopellaar et al. (51), observed that strains of S. mutans which lacked the ability to form water insoluble glucans were noncariogenic in rodents. Recently it has been observed that most strains of S. mutans produce dextranase which is capable of breaking down α-1,6 linked glucan to oligosaccharides. It is postulated that these oligosaccharides are enzymically inserted into the α-1,6 glucan to form complex glucans (52). It has been further shown by Tanzer et al. (53) that strains of S. mutans which lacked dextranase are minimally cariogenic in rodents.

The precise role of extracellular polysaccharide in the pathogenesis of dental caries remains unclear. There is little doubt that many microorganisms in dental plaque have the ability to metabolize glucan and fructan to acid. It also appears that glucan enhances the ability of microorganisms to colonize the tooth, probably protects bacteria from inimical influences, and almost certainly limits the diffusion of ionic substances into and out of plaque, thereby preventing the rapid neutralization of acids formed within dental plaque (54).

Many microorganisms in dental plaque have the ability to form intracellular polysaccharide from a wide variety of carbohydrates (55). As judged by its properties, this appears to be glycogen and is readily catabolized. S. mutans mutants lacking the ability to synthesize glycogen are less cariogenic than are parent wild type strains (56).

The mere physical presence of plaque on the tooth surface does not give rise to dental caries. Pathogenic plaque has considerable capacity to form acid rapidly from a wide variety of carbohydrates. For example, following exposure of plaque to sugar solutions pH values as low as 4.5 are not uncommonly observed; enamel starts to dissolve rapidly at pH 5.5 (57). Dental

caries does not develop in the absence of dietary substrate; rats which receive their diet by gastric intubation remained caries free (58). Plaque in monkeys
which received their diet by a similar route is nonacid-
ogenic; however, plaque which formed in the presence of
glucose or sucrose has considerable acidogenic potential
(59).

The ability of plaque to lower the pH of sugar
solutions is influenced by its age. Older plaque
appears to be much more acidogenic than young plaque,
and the maximum ability to produce acid appears to be
reached in 72-hour plaque. In addition, it appears that
the time taken for older plaque to return to neutral pH
is greater than that observed with young plaque (60).

Although it is apparent that acid production by
dental plaque plays an essential role in the pathogenesis of dental caries, relatively little attention has
been given to the types of acids produced in dental
plaque. "Fasting" plaque was observed by Geddes to
contain 3×10^{-5} mmoles of acid/mg wet weight and that
approximately 5 minutes after exposure to sugar the
concentration increased to 5×10^{-5}. The major change
was a 5-fold increase in D-lactate and an 8-fold in
D + lactate (61).

It is frequently assumed that lactic is the only,
or at least, the predominant acid produced in dental
plaque. However, it has been observed by Gilmour and
Poole (62) that the pH decrease in plaque was not related to concentrations of acetic, propionic or lactic
acids alone and, in addition, they observed that the
concentration of lactic acid was similar to or less than
that of either acetic or propionic. In addition, lactic
acid never exceeded 47% of the total concentration of
the assayed acids, and in only 8% of the plaques examined did it constitute the major acid. Acetic and formic were the major products in 78% and 14% of the plaques
studied. However, the bacterial composition of the
plaques studied was not established. Hence it is not
certain that the plaques studied were, indeed, cariogenic ones.

Sugar alcohols, such as xylitol, sorbitol and
mannitol are not readily fermented by plaque (63,64,65)
and 2 years after continuous exposure in primates little

adaptation of the microbial ecology occurs even though
significant levels of plaque are formed. The acid products of the substances in plaque apparently have not
been determined. Dallmeier et al. (66), using pure
cultures of streptococci have shown that the main end
products when sorbitol is fermented are formic acid and
ethanol. These observations explain in part the relative lack of cariogenicity of xylitol (67), sorbitol and
mannitol (68).

It is clear that dental caries represents a major
public health problem and that it is far more expedient
to attempt to prevent the disease than to treat it.
Fundamentally, three approaches are available to prevent the disease. The tooth can be rendered more
resistant to carious attack; the microbial agent may be
modified to reduce or eliminate the pathogenic capacity
of plaque, and readily fermentable carbohydrates can be
removed from the diet.

Water fluoridation is by far the most effective
public health measure available for the prevention of
caries. Approximately 50% of the population of the
United States now consumes water fluoridated at approximately 1 ppm. The procedure reduces the incidence of
caries by 50-60% (69). In some areas, however, it is
not possible to fluoridate municipal water supplies and,
therefore, alternative methods of providing fluoride
must be used. It has been observed that where school
water supplies are fluoridated at 5 to 7 times optimum
that a reduction of 35 to 40% in the incidence of caries
is achieved (70). The use of school fluoride mouthrins-
ing programs, too, have been shown to be effective (70,
71). The conscientious use of fluoride dentifrices
results in a reduction of 15 to 20% in the incidence of
caries in children (71). It is generally agreed that
if a method could be devised which would permit low
levels of fluoride to be present continuously in the
mouth, that substantial benefit could accrue. Such
devices have been fabricated and are being actively
researched (72).

The use of antimicrobial agents to prevent dental
caries has attracted attention since the role of microorganisms in dental caries has been recognized. Thus-
far an agent suitable for prolonged use has not been

identified. However, chlorhexidine illustrates the
type of agent which is likely to be successful. Chlorhexidine has considerable substantivity and thus remains
in the mouth for prolonged periods (73). It has substantial plaque-restricting properties, however, results
from caries trials are equivocal (74). It seems probable that by using slow release devices, agents which
are presently ineffective because they are cleared from
the mouth rapidly could be rendered therapeutically
effective.

Following the identification of specific microorganisms capable of inducing dental caries, the possibility of developing a vaccine against caries has become
increasingly attractive. A number of investigators
have reported considerable success in preventing caries
in rodents by immunization against S. mutans (75,76,77);
others have either had equivocal results or failure. In
a study carried out over 5 years, Bowen et al. (78),
observed that primates vaccinated with broken S. mutans
intramucosally remained virtually caries-free. In
contrast, primates vaccinated using a glucosyltransferase preparation developed rampant caries.

The immunoglobulin class which contains the protective antibody appears to vary from species to species.
For examples, in primates that protective antibody
appears to be located in IgG (70), whereas in contrast
Taubman et al. (80), and McGhee et al. (81), reported
that protective antibody in rodents appears to reside
primarily in the IgA class. Clearly, the most appropriate antigen, the optimum dose, and route of administration needs to be determined before a vaccine
suitable for investigation in humans can be available.

There is clear evidence that the level of dental
caries is related to frequency of ingestion of readily
fermentable carbohydrate (82). The benefits which may
be derived from curbing the frequency of intake of
carbohydrate can be seen from the results of at least
two investigations. It was observed by Gustafson et
al. (83) that patients who consumed 95 kg. of sugar per
year with meals developed 4-5 times fewer carious
lesions than those who ingested 85 kg., 15 of which was
taken between meals. Similar results were achieved by
Harris et al. (84), who observed that children residing
in an institution who had restricted access to sugar

and sugar-containing products had substantially fewer lesions than comparable non-institutionalized children.

Although restriction of the intake of sugars has been shown to be an effective method for prevention of dental caries, the public seems unwilling to accept this approach on a general basis, largely perhaps because eating candy, which is a pleasant experience, is not readily associated with the development of carious lesions at some future time.

Much research has been carried out to identify alternatives to sucrose. The sugar alcohols xylitol, sorbitol and mannitol are either noncariogenic or minimally cariogenic. Claims that xylitol is cariostatic (85) remain to be substantiated. These substances are available in sugarless gums and soft drinks usually in combination with saccharin. Many other sweet tasting compounds, e.g., aspartame, dihydrochalcones are probably noncariogenic but are not yet available to the public.

It is apparent that many approaches to the prevention of dental caries are available, and others are under active investigation, however, because of the prevalence of the disease only public health measures are likely to have great impact and, of course, water fluoridation is presently the most effective.

I.   REFERENCES

1. Scherp, H.W., Science 173, 1199 (1971).
2. Dawes, C., Jenkins, G.N., and Tonge, C.H., Br. Dent. J. 115, 65 (1963).
3. Leach, S.A., in "Dental Plaque" (W.D. McHugh, Ed.), p.143, E & S Livingstone Ltd., Edinburgh (1970).
4. Jenkins, G.N., in "Prevention of Periodontal Disease" (J.E. Eastoe, D.C. Picton, Eds.), p.34, Henry Kimpton, London (1971).
5. Schroeder, H.E., in "Formation and Inhibition of Dental Calculus" (Hans Huber, Ed.) Berne, Switzerland (1969).
6. Tatevossian, A., and Gould, C.T., Arch. Oral Biol. 21, 313 (1976).

7. Tatevossian, A., and Gould, C.T., Arch. Oral Biol. 21, 319 (1976).
8. Holt, R.L., and Mestecky, J., J. Oral Pathol. 4, 86 (1975).
9. Taubman, M.A., Arch. Oral Biol. 19, 439 (1974).
10. Guggenheim, B., and Schroeder, H.E., Helv. Odontol. Acta 11, 131 (1967).
11. Critchley, P., Saxton, C.A., and Bowen, W.H., IADR (Brit. Div.) Abstract #157 (1972).
12. Wood, J.M., and Critchley, P., Arch. Oral Biol. 11, 1039 (1966).
13. Critchley, P., Caries Res. 3, 249 (1969).
14. Gold, W., Preston, F.B., and Blechman, H., J. Periodontal Res. 44, 263 (1973).
15. Hotz, P., Guggenheim, B., and Schmid, R., Caries Res. 6, 103 (1972).
16. McDougall, W.A., Aust. Dent. J. 9, 1 (1964).
17. van Houte, J., and Jansen, H.M., Caries Res. 2, 47 (1968).
18. Ashley, F.P., Arch. Oral Biol. 20, 167 (1975).
19. Streckfuss, J.L., Smith, W.N., Brown, L.R., and Campbell, J., J. Bacteriol. 120, 502 (1974).
20. Bowen, W.H., and Cornick, D., Int. Dent. J. 20, 382 (1970).
21. Birkeland, J.M., Caries Res. 4, 243 (1970).
22. Hardwick, J.L., in "Dental Plaque" (W.D. McHugh, Ed.), p.171, E & S Livingstone Ltd., London and Edinburgh (1970).
23. Miller, W.D., Dent. Cosmos 44, 425 (1902).
24. Clarke, J.K., Br. J. Exp. Pathol. 5, 141 (1924).
25. Fish, E.W., and McClean, I.H., Dent. Cosmos 76, 837 (1934).
26. McClean, I.H., Br. Dent. J. 48, 579 (1927).
27. Stephan, R.M., J. Am. Dent. Assoc. 27, 718 (1940).
28. Jay, P., Am. J. Orthod. 33, 162 (1947).
29. Hill, T.J., J. Am. Dent. Assoc. 26, 239 (1939).
30. Keyes, P.H., Arch. Oral Biol. 1, 304 (1960).
31. Fitzgerald, R.J., and Keyes, P.H., J. Am. Dent. Assoc. 61, 9 (1960).
32. Gibbons, R.J., Berman, K.S., Knoettner, P., and Kapsimalis, P., Arch. Oral Biol. 11, 549 (1966).
33. Krasse, B., Arch. Oral Biol. 10, 223 (1965).

34. Zinner, D., Jablon, J., Aran, A., and Saslaw, M.,
    Proc. Soc. Exp. Biol. 118, 766 (1965).
35. Bowen, W.H., Caries Res. 3, 227 (1969).
36. Klein, H., J. Am. Dent. Assoc. 33, 735 (1946).
37. Böök, J.A., and Grahnen, H., Odontol. Revy. 4, 1
    (1953).
38. Davies, G., in "Caries Resistant Teeth" (G.
    Wolstenholme and M. O'Connor, Eds.), p.7, J and A
    Curchill, London (1965).
39. Berkowitz, R.J., Jordan, H.V., and White, G.,
    Arch. Oral Biol. 20, 171 (1975).
40. Carlsson, J., Soderholm, G., and Almfeldt, I.,
    Arch. Oral Biol. 14, 243 (1969).
41. Cornick, D., and Bowen, W.H., Br. Dent. J. 130,
    231 (1971).
42. van Houte, J., and Duchin, S., Arch. Oral Biol.
    20, 771 (1975).
43. Sonju, T., Christensen, T.B., Kornstad, L., and
    Rolla, G., Caries Res. 8, 113 (1974).
44. Krogstad, S., Sonju, T., Melsen, B., and Rolla, G.,
    J. Biol. Buccale 3, 53 (1975).
45. Markham, J., Knox, K., Wicken, A., and Hewett,
    M.S., Infect. Immun. 12, 378 (1975).
46. Kelstrup, J., and Funder-Nielsen, T.D., Arch. Oral
    Biol. 17, 1659 (1972).
47. Hay, D.I., Gibbons, R.J., and Spinell, D.M.,
    Caries Res. 6, 111 (1971).
48. Ericson, T., Carlen, A., and Dagerskagt, in
    "Microbial Aspects of Dental Caries" (H.M. Stiles,
    W.J. Loesche and T.C. O'Brien, Eds.), p.151,
    Information Retrieval, Inc., Washington, D.C. and
    London (1977).
49. Rölla, G., and Kilian, M., Caries Res. 11, 85
    (1977).
50. Mukerjee, P., Adv. Colloid Interface Sci. VI, 24,
    (1967)
51. deStoppelaar, J.O., König, K.G., Plasschaert,
    A.J.M., and vander Hoeven, J.S., Arch. Oral Biol.
    16, 971 (1971).
52. Schachtele, C., Harlander, S., and Germaine, G.,
    J. Dent. Res. 56 (Spec. Issue) A133 (1977).
53. Tanzer, J.M., Freedman, M., and Rinehimer, L.A.,
    J. Dent. Res. 56 (Spec. Issue) A133 (1977).

54. Critchley, P., Caries Res. 3, 249 (1969).
55. van Houte, J., Winkler, K.C., and Jansen, H.M.,
    Arch. Oral Biol. 14, 45 (1969).
56. Tanzer, J.M., Freedman, M.L., Woodiel, F.M.,
    Eifert, R.L., and Rinehimer, L.A., in "Microbial
    Aspects of Dental Caries" (H.M. Stiles, W.L.
    Loesche and T.C. O'Brien, Eds.), p.597, Informa-
    tion Retrieval, Inc., Washington, D.C. and
    London (1977).
57. Kleinberg, I., and Jenkins, G.N., Arch. Oral Biol.
    9, 493 (1964).
58. Kite, O., Shaw, J.H., and Sognnaes, R.F., J. Nutr.
    42, 89 (1950).
59. Bowen, W.H., Arch. Oral Biol 19, 231 (1974).
60. Graf, H., Schweiz, Mschv. Zahnheiek 79, 146 (1969).
61. Geddes, D.A., Arch. Oral Biol. 17, 537 (1972).
62. Gilmour, M.N., and Poole, A.E., Caries Res. 1, 247
    (1967).
63. Cornick, D., and Bowen, W.H., Arch. Oral Biol. 17,
    1637 (1972).
64. Makinen, K., and Scheinin, A., Int. Dent. J. 21,
    331 (1971).
65. Makinen, K., and Scheinin, A., Acta Odontol. Scand.
    30, 259 (1972).
66. Dallmeier, E., Bestman, H.J., and Kronche, A.,
    Dtsch Zahnaerztl. 25, 887 (1970).
67. Scheinen, A., Makinen, K., and Ylitalop, Acta
    Odontol. Scand. 32, 383 (1974).
68. Ahlden, M.L., and Frostell, G., Odontol. Revy. 26,
    1 (1975).
69. Ericson, S.Y., Caries Res. 11 (Suppl 1) 2 (1977).
70. Horowitz, H.S., Creighton, W.E., and McClendons
    Arch. Oral Biol. 16, 609 (1971).
71. Horowitz, H.S., Heifetz, S.B., and Law, F.E.,
    J. Am. Dent. Assoc. 88, 352 (1974).
72. Mirth, D.B., and Bowen, W.H., in "Microbial
    Aspects of Dental Caries" (H.M. Stiles, W.J.
    Loesche and T.C. O'Brien, Eds.), p.249, Informa-
    tion Retrieval, Inc., Washington, D.C. and
    London (1977).
73. Rölla, G., Löe, H., and Schiøtt, R., Arch. Oral
    Biol. 16, 1109 (1971).

74. Eriksen, H.M., Gjermo, P., and Johansen, J.R.,
    Helv. Odontol. Acta 17, 52 (1973).
75. Tanzer, J.M., Hageage, G.J., and Larson, R.,
    Arch. Oral Biol. 18, 1425 (1973).
76. McGhee, J.R., Michalek, S.M., Webb, J., Navia,
    J.M., Rahman, A.F., and Legler, D., J. Immunol.
    114, 300 (1975).
77. Guggenheim, B., Muhlemann, H.R., Regolati, B., and
    Schmid, R., in "Dental Plaque" (W.D. McHugh, Ed.),
    p.287, E & S Livingstone, London and Edinburgh
    (1970).
78. Bowen, W.H., Cohen, B., Cole, M.F., and Colman, G.,
    Br. Dent. J. 139, 45 (1975).
79. Lehner, T., Challacombe, S., and Caldwell, J.,
    J. Dent. Res. 55, 166 (1976).
80. Taubman, M.A., and Smith, D.J., Infect. Immun. 9,
    1079 (1974).
81. Michalek, S., McGhee, J., Mestecky, J., Arnold, R.,
    and Bozzo, L., Science 192, 1238 (1976).
82. Zita, A., McDonald, R.E., and Andrews, A.L.,
    J. Dent. Res. 38, 860 (1959).
83. Gustafson, B.E., Quensel, C.E., Swedanev, L.,
    Lundquist, C., Grahnen, H., Bonow, B., and Krasse,
    B., Acta Odontol. Scand. 11, 232 (1954).
84. Harris, R.M., J. Dent. Res. 42, 1387 (1963).
85. Scheinin, A., Makinen, K.K., Tammisalo, E., and
    Rekola, S., Acta Odontol. Scand. 33, 2369 (1975).

DISSOLUTION MECHANISMS OF THE APATITE CRYSTALS DURING DENTAL CARIES AND BONE RESORPTION

R.M. Frank
J.C. Voegel

Groupe de Recherches Inserm U 157, Faculté de Chirurgie Dentaire, Université Louis Pasteur, STRASBOURG

High resolution electron microscopy, electron diffraction and chemical kinetic techniques were used to study the dissolution of tooth and bone crystals. Width and thickness of enamel, dentin and bone apatite crystals were determined for crystals presenting simultaneously (100) and equivalent fringe patterns. The mean width to thickness ratios for enamel, dentin and bone crystals were respectively 1.99 ± 0.10, 3.60 ± 0.14 and 7.43 ± 0.27, indicating a clear trend from a hexagonal prismatic shape to a true platelet configuration. Great similarity was observed in the mechanisms of crystal dissolution in dental caries and bone resorption. With increasing width to thickness ratios, one or several central holes developed along the $\underline{c}$ axis with fast kinetics. The extensions of the central holes along (100) planes in dentin and bone crystals are very similar to the lateral dissolution steps with slow kinetics observed in enamel crystals. Through single crystal electron diffraction studies, a significant increase in the $\underline{a} = \underline{b}$ parameters of 0.05 Å was observed in the carious enamel crystals by comparison with non-carious enamel. These different observations indicate that the biological apatite crystal dissolution mechanisms which occur during

dental caries and bone resorption are highly syste-
matized and follow very precise rules and that there
exist preferential dissolution sites in apatite
crystals.

I. INTRODUCTION

All calcified tissues (bone and tooth) share a common inorganic crystalline phase similar to hydroxyapatite: $Ca_{10}(PO_4)_6(OH)_2$. Few mineral substances have
been the subject of so much investigation and controver-
sy among chemists, physical chemists, crystallographers,
and biologists. By high resolution transmission electron microscopy, it has been possible to demonstrate
periodic lattice images in apatite crystals of human
bone, cementum, dentin and enamel (1,2,3,4). Since the
represented crystal planes are nearly parallel to the
electron beam, the fringes obtained can be considered as
the projections of the different planes. These projec-
ted periodic lattice images allow the determination of
the precise size and shape of their monocrystals in relation to their crystallographic axes (5). Voids and
imperfections in the crystal lattice can also be visu-
alized (6).

We have applied this technique of high resolution
transmission electron microscopy to the determination
of the shape and size of human enamel, dentin and bone
crystals as well as to the study of the dissolution mechanisms of the apatite crystals during dental caries
and bone resorption. These investigations were accompanied by kinetic studies of enamel crystal dissolution
in buffered acid solutions as well as by electron diffraction investigations of single non-carious and carious enamel crystals.

II. MATERIALS AND METHODS

So-called enamel white spots and adjacent normal
human enamel were obtained from permanent molars and
premolars of 14-30 year old patients. Samples of carious dentin were taken from the coronal part of molars

and premolars. The dentin underlying fissure carious
lesions was sampled from the opaque dentin layer. Adjacent normal dentin was also collected. Human alveolar
bone in advanced periodontitis cases was obtained from
50-65 year old patients with pocket depths of 5 mm or
more, advanced tooth mobility, and radiographically visible bone loss. The bone fragments were collected immediately after tooth extraction. Adjacent normal bone
was also sampled.

   After fixation for 3 hr in 2%   glutaraldehyde-
paraformaldehyde in 0.1 M cacodylate buffer at pH 7.4
and for 1 hr in 2% osmic acid in the same buffer, the
specimens were Epon embedded and cut with a Sorvall-
Porter microtome equipped with a diamond knife. The
sections were examined in a Jeol 100 B transmission
electron microscope using a "cold beam" at 100 KV with
a specimen anticontamination device. The equidistances
of the different family planes were assessed by determination of the mean value of all the equivalent fringes
in the crystal lattice.

   The single enamel crystals used for electron diffraction were selected under magnification of 150,000
diameters. In carious enamel, only those apatite crystals presenting a central core lesion along the c axis
were diffracted. Calibration of the electron microscope
was achieved by comparison with electron diffraction
patterns of evaporated lithium chloride and thallium
chloride: 75 single crystal diffraction patterns of
normal enamel were compared with the same number of analogous carious crystal patterns according to the technical procedure detailed by Voegel and Frank (7), using a
top entry goniometer.

   The chemical kinetic studies of crystal dissolution
utilized 100-200 mesh synthetic hydroxyapatite powder
(Bio-Rad, Richmond, California) and human enamel powder:
100 mg of apatite were placed in 100 ml of sodium
citrate-citric acid buffer solution at either pH 3.7 or
5.6. Variable amounts of sodium fluoride (from 0 to 15
ppm of F^-) were added. Calcium determinations on 100
μl of solution collected from time zero to 24 hr were
carried out with a Beckman atomic absorption spectro-
photometer after addition of lanthanum oxide for decom-
plexation of all ionic calcium. Samples of synthetic

hydroxyapatite exposed to the buffered acid solutions
were collected at various time intervals for examination
in the electron microscope.

III. SIZES AND SHAPES OF ENAMEL, DENTIN AND BONE APÁ-
      TITE CRYSTALS

     From a survey of the literature, it appears that
little agreement has been reached among investigators
about dimensions and shape of the biological apatite
crystals. Line broadening measurements of x-ray diffraction patterns and, more often, direct estimations
of transmitted electron micrographs have been used.
Enamel apatite has been recognized as being the largest
and longest biological crystal. It has been described
as a rod-shaped crystal (8), as an elongated flattened
hexagon (9,10), as an irregular hexagon (11), as a
prismatic hexagon (12,13) and as a rectangle (14).
Dentin and bone crystals have been considered either as
needle-shaped or as true platelets. However, Johansen
and Parks (15) and Johansen (16) by tilting specimens
under the electron beam observed that the needles appear plate-like.
     We have applied high resolution electron microscopy
to the size determinations of the biological apatites.
The presence of two sets of equidistant and parallel
fringes in a monocrystal allows exact measurements.
Figures 1,2, and 3 are presented here to orient the
reader with the general form of apatite crystals. The
evidence for the drawing of these diagrammatic repre-
sentations as shown will be presented subsequently in
this report.
     We took into account only such crystals presenting
simultaneously (100) planes (Fig. 2,3) and equivalent
fringe patterns. The hexagonal fringe patterns require
that all of the three lattice plane sets equivalent to
the (100) planes are nearly parallel to the electron
beam or, in other words, that the c axis is perpendicular to the image (Fig. 4). The width of the monocrystals measured perpendicularly to (100) planes is called
L and the thickness along (100) is called E (Fig. 1).

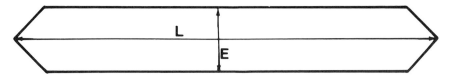

Fig. 1. *Schematic cross-section perpendicular to the c axis of an apatite monocrystal with the indication of the width (L) and the thickness (E).*

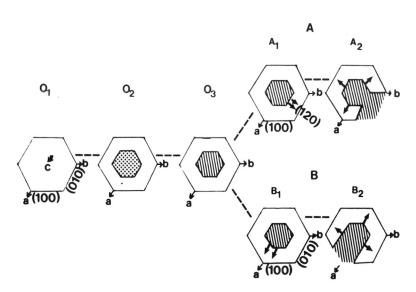

Fig. 2. *The different steps of the enamel monocrystal dissolution during the carious process seen in a direction perpendicular to the c axis.*

It became apparent that great variations in crystal width and thickness existed in the same samples of enamel, dentin and bone. In enamel, for example, true hexagonal outlines were observed (Fig. 5), but flattened hexagonal shapes were also noted (Fig. 4). The variable flattened hexagonal form (Fig. 6) of intertubular dentin became striking on cross sections of such crystals (Fig. 7 and 8). However, similar sections made on bone

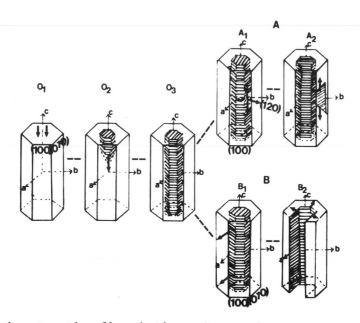

Fig. 3. The dissolution of a carious enamel apatite crystal in a direction parallel to the c axis. After the development of a central core lesion leading to a hollow crystal (O₂, O₃), lateral side dissolutions occur either along (120) planes (mechanism A) or along (010) or (100) planes (mechanism B).

crystals showed even more flattened hexagonal contours (Fig. 9,10,11), implying clearly a platelet configuration (Fig. 12).

The measurements of width (L) and thickness (E) on cross-sections of crystals made exactly perpendicular to the c axis, with presence of two sets of equidistant and parallel fringes (Fig. 4,8) allowed precise estimations of L/E ratios (Table I). Striking differences were noticed for the three types of crystals. Whereas the enamel crystals showed the highest mean width and thickness, the variations of monocrystal size and L/E ratios were clearly more variable for dentin (Fig. 14) than for enamel (Fig. 13) and bone (Fig. 15). For enamel, a slightly flattened hexagonal shape was found with a mean L/E ratio of 1.99. Jongebloed (10) reported

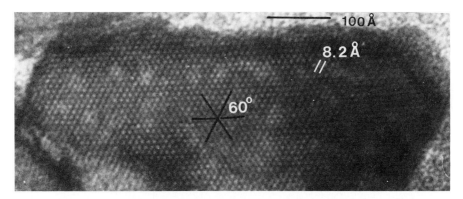

Fig. 4. Transverse section perpendicular to the c axis of an intact human enamel crystal with flattened hexagonal contours. Three series of fringes, forming 60° angles between them, correspond to the (100) family planes, with 8.2 Å equidistances.   x 1,980,000

a higher width to thickness average ratio (2.4 ± 0.3), whereas Selvig and Halse (17) indicated for mature rat enamel a ratio of 1.82. In dentin, a range of L/E values from 1 to 7 was found, whereas values lower than 3 were never found for bone monocrystals, indicating their platelet configuration. Thus, gradation from slightly flattened hexagonal prism form toward thin platelet shape was apparent from enamel to dentin bone.

The total length of the crystals is difficult to assess on thin sections and average length for enamel crystals of 5000 Å (18) and 6000 Å (12) have been reported. Enamel crystals longer than 2000 Å have been shown with darkfield electron microscopy (19). For bone, lengths in the range of 200-700 Å have been reported (5,20). The general shape of the bone crystal is a lengthened and slightly bent hexagon with a very large radius of curvature.

Although the shape and size of such crystals are variable, it has been shown through high resolution microscopy that the basic structures are identical. Periodic parallel fringes and rhombohedral or hexagonal patterns have been demonstrated with equidistances

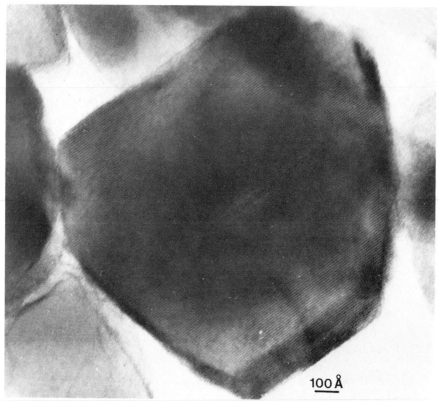

100 Å

Fig. 5. Cross-section perpendicular to the c axis of an intact human enamel crystal with typical hexagonal contours with 8.2 Å equidistant fringes corresponding to (100) planes. x 750,000

ranging from 3.4 Å to 18 Å (1,2,3,4,5,9,21). A periodic lattice image of 8.2 Å, corresponding to the resolution of (100) planes of the hydroxyapatite unit cell is visible in enamel (Fig. 4), dentin (Fig. 8), and bone (Fig. 9,10).

High resolution electron microscopy allows the direct visualization of certain crystalline defects such as dislocations which have been demonstrated in biolo-

DENTINE

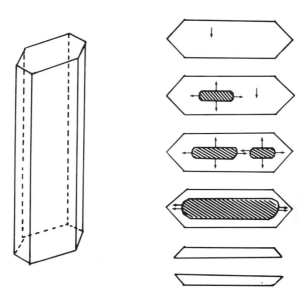

Fig. 6.   Theoretical dentin monocrystal with a flattened hexagonal prismatic form (left side).   On the right side, from top to bottom, different dissolution stages of the dentin monocrystal during the carious process, on a transverse section made perpendicularly to the c axis.

gical apatites, especially in the central regions of the enamel crystals (2,6,22), as well as in synthetic apatite monocrystals (10,23).   With this technique, it is, however, not possible to differentiate between screw dislocations and edge dislocations.   We have observed a greater density of dislocations in bone crystals than in dentin and enamel crystals.

IV.   DISSOLUTION MECHANISM OF ENAMEL AND DENTIN APATITE
      CRYSTALS DURING DENTAL CARIES

Enamel and dentin caries develop in close relationship to the micro-organisms of the dental plaque (Fig. 16,17).   By microradiography, it is classically conclu-

*Fig. 7.   Cross-section perpendicular to the c̲ axis
of a dentin monocrystal with a L/E ratio of 2.97.
x 1,140,000*

TABLE I   Range and Mean Value in Width (L)
and Thickness (E) as well as L/E Ratios
of Human Enamel (A), Dentin (B)
and Bone (C) Apatite Crystals

		Width (L)	Thickness (E)	L/E
A	Range:	539 - 1515 Å	216 - 1067 Å	1.99 ± 0.10
	Mean:	890.80 Å ± 25.80	503.60 Å ± 20.80	
B	Range:	100 - 900 Å	40 - 170 Å	3.60 ± 0.14
	Mean:	364.50 Å ± 14.50	103.30 Å ± 2.70	
C	Range:	270 - 900 Å	40 - 130 Å	7.43 ± 0.27
	Mean:	562.10 Å ± 19.00	79.10 Å ± 3.10	

Fig. 8. Cross-section perpendicular to the c̲ axis of a dentin crystal having a L/E ratio of 4.37. Three series of fringes corresponding to (100) planes and equivalent planes are visible.   x 1,690,000

ded that the so-called caries white spot consists of an apparently intact surface layer overlying a sub-surface demineralization. However, when using non-decalcified thin sections, four types of ultrastructural lesions can be located in the superficial enamel layer which is more often of the aprismatic type (19). Beside a scattered apatite crystal dissolution in close vicinity of the dental plaque (Fig. 16), the enamel surface can also be indented by irregular destruction of the superficial apatite crystals. A localized hemispheric surface dissolution of the crystals has also been noted. Finally, small microdefects, 0.2 to 2 microns large, starting at

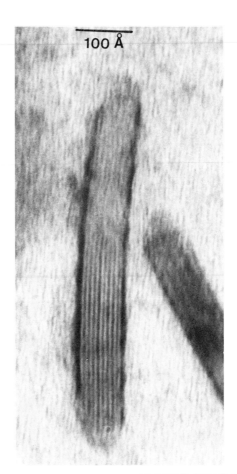

Fig. 9.   Cross-section of a slightly bent bone monocrystal showing only one set of fringes.   L/E ratio = 8.   x 1,490,000

the enamel surface, reach the deeper enamel layer.   In these defects, an amorphous organic stroma has been observed replacing the dissolved crystals.

Good agreement has been reached on the carious destruction of prismatic enamel.   The first apparent phenomenon is a broadening of the interface between rod and interrod substance (19).   A scattered apatite crystal dissolution is then observed in the rods, whereas, at this stage the crystals of the interrod substance

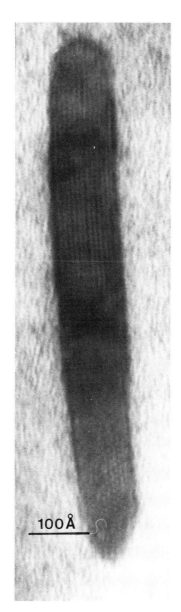

*Fig. 10.   Cross-section of a bone monocrystal per-
pendicular to the c̲ axis.   L/E ratio of 7.36.
x 1,630,000*

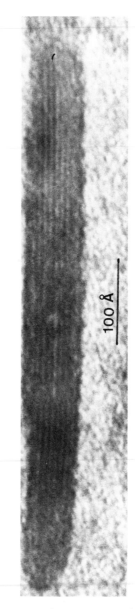

*Fig. 11.   Cross-section perpendicular to the c̲ axis
of a bone monocrystal with a L/E ratio of 9.18.
x 2,250,000*

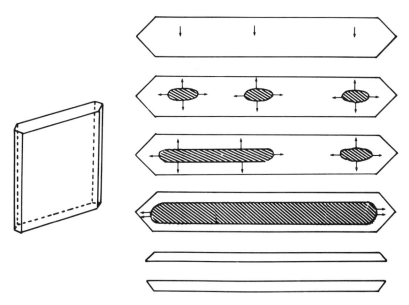

Fig. 12. Schematic representation of apatite bone crystal with its platelet form. On the right side, various stages of dissolution during bone resorption are shown in a monocrystal cross-section perpendicular to the c axis.

are more or less spared (Fig. 18). After demineralization of the prism core, the crystals of the interrod substance are also destroyed.

The dissolution mechanism of the individual enamel apatite crystal during enamel caries both in the surface and deeper layers is basically identical. It is characteristically a highly systematic process following exactly the crystallographic planes (21,22). Enamel monocrystal dissolution starts always at one end of the crystal through formation of a central core lesion, with hexagonal outlines, which extend anisotropically along the c axis (Fig. 2,3,19,20). Along the a and b axes, this central hole follows exactly the $(1\bar{0}0)$ and $\bar{(010)}$ planes, forming between them a 60° angle (Fig. 21). According to Arends et al. (23) and Jongebloed (10), the development of this central core dissolution is facilitated by the presence of helicoidal defects along the c axis. We confirm this point of view because of

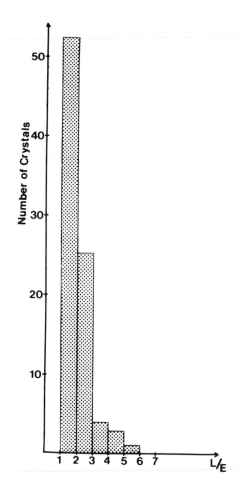

Fig. 13.   L/E ratio distribution of the human ena-
mel monocrystals with a mean value of 1.99 ± 0.10.

1) the fast and anisotropic development of the crystalline destruction along the c̲ axis, 2) the dissolution
development parallel to the outer hexagonal crystalline
faces, and 3) the presence of dislocations (22). It
must also be remembered that screw dislocations are the
basis of the helicoidal growth theory of the monocrystals.
        This central dissolution leading to a hollow crys-
tal reaches a maximum width of 60-70 Å. Along a̲ and b̲

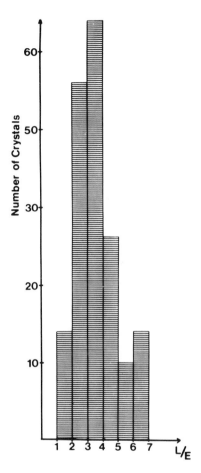

Fig. 14.   L/E ratio distribution of the human dentin monocrystals with a mean value of 3.60 ± 0.14.

axes the central lesion stops after dissolution of about 8-(100) unit cell planes.

After this initial step of the central core lesion, enamel monocrystal dissolution progresses to the development of lateral dissolution (Fig. 2 and 3). The most frequent type of lateral dissolution (Fig. 2 and 3,A) is characterized by a crystalline destruction which follows the (120) planes, perpendicular to (100) and (010) planes. In cross sections, it consists in a conical defect with a base corresponding to the mono-

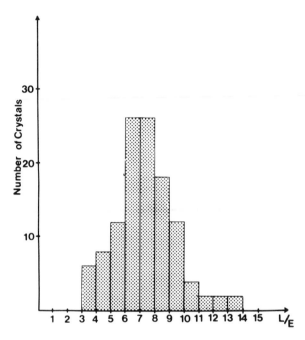

Fig. 15.   L/E ratio distribution of the human bone monocrystals with a mean value of 7.43 ± 0.27.

crystal surface (Fig. 22).   In the second type of lateral dissolution (Fig. 2 and 3,B), crystal destruction
develops along (010) and (100) planes parallel to the
a and b axes (Fig. 23).   Both types of lateral dissolution can sometimes be observed in the same monocrystal
(Fig. 24).
    The fragmentation of the hollow crystal through
lateral dissolution extending along a and b axes can
occur unilaterally or on different sides.   The total
destruction of the enamel crystal results in a lateral
extension of dissolution perpendicular or parallel to
(100) planes.   It is interesting that this process of
dissolution during dental caries is similar to the
destruction of synthetic hydroxy or fluoroapatites

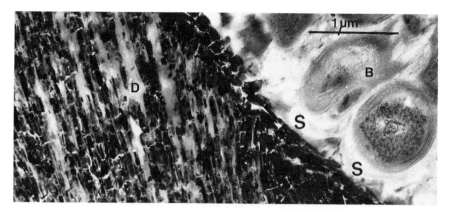

Fig. 16.   Scattered apatite crystal dissolution in the enamel surface layer of a fissure carious lesion. D = areas of complete apatite crystal dissolution. S = enamel surface.   B = micro-organism of the dental plaque.   x 27,000

after treatment with lactic or citric acid, occurring in relation to the presence of central screw dislocations along the c axis (10,23).

The behavior of the dentin apatite crystal was followed in the intertubular substance of the so-called opaque carious dentin (Fig. 17). In dissolving mono-crystals, having a width to thickness ratio lower than two (Fig. 25), only one central hole lesion was ob-served. With increasing L/E ratios, one (Fig. 26), two or several dissolution centers (Fig. 6,27) appeared and developed on (001) planes along the c axis. The central hole often had an hexagonal outline (Fig. 26). When the L/E ratio was higher than 5, rows of central hole lesions progressively appeared (Fig. 27). The extension of the central core lesions along (100) planes in the direction of the crystal width as well as an extension along the c axis leads to complete destruction of the central part of the dentin monocrystal which is then reduced to two thin plates (Fig. 6,28).

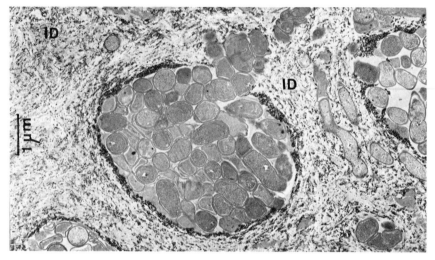

*Fig. 17.   Transverse section of carious dentinal
tubules filled with numerous bacteria and circumscribed
by a thin ring of electron dense inorganic material.
There is a break in these remnants of the peritubular
dentin and micro-organisms are observed in the demineralized intertubular dentin (ID).   x 11,000*

V.   DISSOLUTION MECHANISM OF APATITE CRYSTALS DURING BONE RESORPTION

Bone resorption along the human alveolar bone surface was studied in cases of advanced periodontitis.
The dissolution of the bone monocrystals was observed
in the 3-4 micron wide borders of Howship lacunae in
close contact to the ruffled border of the osteoclast

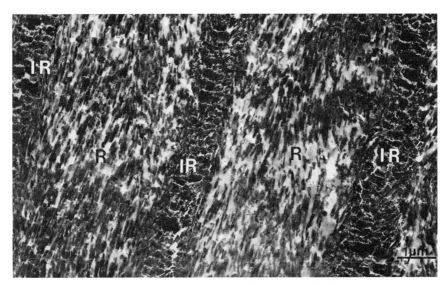

*Fig. 18. The apatite crystal dissolution in cari-
ous human enamel. Dissolution is more advanced in the
rods (R) than in the interrod substance (IR). x 12,000*

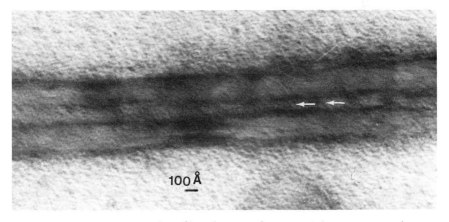

*Fig. 19. Longitudinal sections of human apatite
crystal of carious enamel. A central core lesion (white
arrows) is located along the c axis. x 370,000*

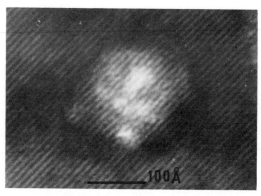

Fig. 20. Hexagonal outlines of a central hole lesion on a cross-section perpendicular to the c axis of a carious enamel crystal.   x 1,610,000

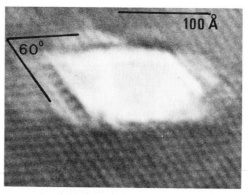

Fig. 21.   Central core lesion of a cross-sectioned enamel apatite crystal.   Along the a and b axes, this central dissolution follows exactly the (Ī00) and (010) planes, forming a 60° angle, with fringe equidistances of 8.2 Å.   x 2,500,000

(Fig. 29). Contrary to the opinion of Bonucci (24), who thought that the solubilization of inorganic crystals does not occur outside the cell, the different dissolution steps were followed in extracellular location on cross-sections of monocrystals (Fig. 29,30). We often observed numerous micro-organisms along resor-

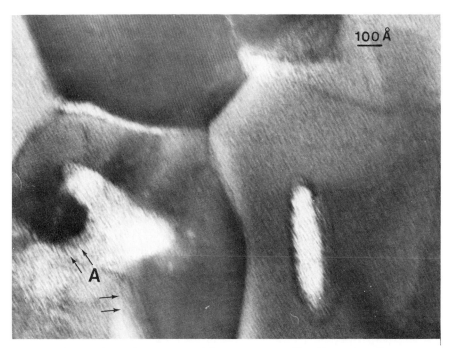

Fig. 22. Cross-section perpendicular to the c axis of several apatite crystals in carious enamel. A central core lesion is visible in the lower right crystal. In the lower left crystal, a central hole is continuous with type A lateral dissolution (arrows). Periodic lattice fringes can be seen in the upper left crystal.   x 72,000

bing bone surfaces, confirming a previous report made by Michel (25). The dissolution mechanism of bone monocrystals adjacent to osteoclasts (Fig. 29) or microorganisms (Fig. 30) is ultrastructurally similar.

As observed in cross-sections perpendicular to the c axis, the initial step of dissolution of the bone crystal was characterized by the development of one or several central core lesions (Fig. 12), sometimes with an hexagonal pattern. These central holes extended in a first stage along the c axis. Through lateral extensions along (100) planes, one or several central clefts

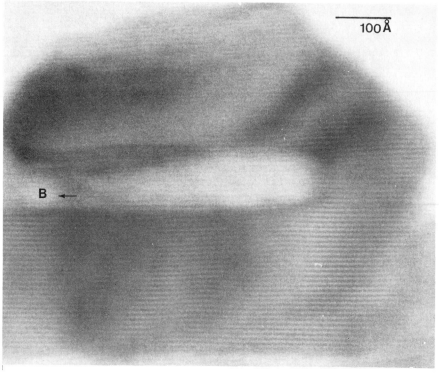

Fig. 23.   Cross-section of a carious human enamel crystal with a central core lesion continuous with type B lateral dissolution.   x 1,680,000

appeared (Fig. 31), which through their fusion (Fig. 12, 32,33), led to a splitting of the bone crystal into two thin plates (Fig. 12,33).

VI. KINETIC STUDY OF THE APATITE CRYSTAL DISSOLUTION

By a combined chemical and electron microscopic study, we have monitored the kinetics of the appearance of the central core lesion and lateral dissolution in synthetic hydroxyapatite as well as in enamel (26). Jongebloed (10) had shown that the central dissolution

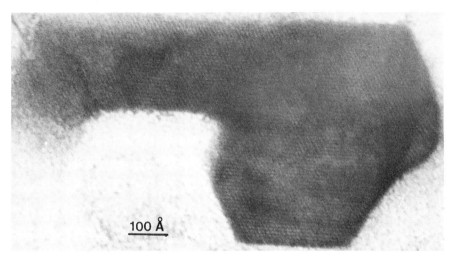

*Fig. 24. Advanced destruction of a flattened he-
xagonal enamel crystal in cross-section through a combination of the two types of lateral dissolution perpendicularly (A) or parallel (B) to (100) planes.
x 1,180,000*

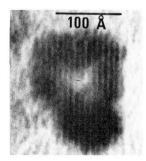

*Fig. 25. Cross-section perpendicular to the c
axis of a carious dentin crystal (L/E ratio = 1.22)
with only one central core lesion.   x 1,850,000*

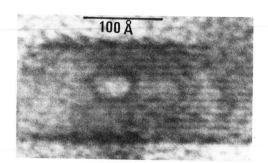

Fig. 26.   A central core lesion with a hexagonal outline visible in cross-section of a carious dentin monocrystal.   x 2,110,000

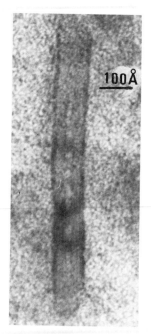

Fig. 27.   Development of several central core lesions along (100) planes in a cross-section of a dentin monocrystal.   x 920,000

Fig. 28. Cross-section of a carious dentin crystal. Confluence of several central core lesions extending laterally along (100) planes apparently leads to central cleft development. x 900,000

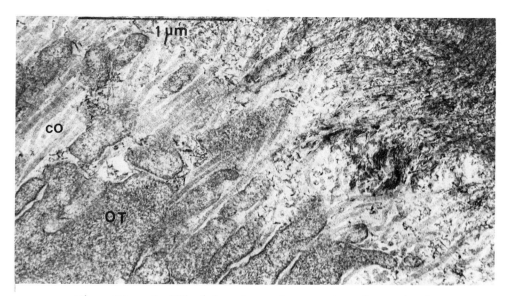

Fig. 29. Ruffled border of an osteoclast (OT) along a resorbing area of human alveolar bone in a case of periodontitis. After disappearance of the apatite crystals, typical collagen fibrils (co) with cross-striations are present. x 43,000

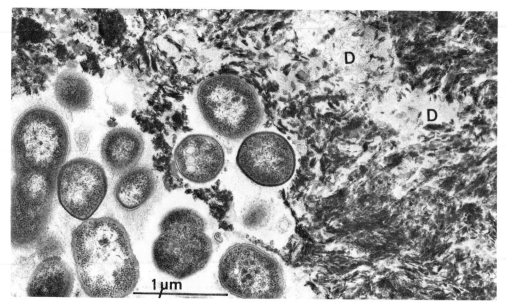

Fig. 30.   Advanced case of periodontitis.   Presence
of numerous micro-organisms along the periodontal surface of the alveolar bone.   Areas of apatite crystal
dissolution (D) are visible.   x 26,000

Fig. 31.   On a cross-section of a bone monocrystal
during bone resorption, several central lesions developing along (100) planes are visible.   x 1,630,000

Fig. 32. Dissolution cleft developing along (100) planes in the center of a cross-sectioned bone monocrystal during bone resorption.   x 710,000

of synthetic hydroxyapatite crystals in 5 M citric acid at 37°C is a fast phenomenon occurring at a velocity of about 600 Å/sec.

By treating synthetic hydroxyapatite and enamel apatite crystals in sodium citrate-citric acid buffer solutions at pH 3.7 (Fig. 34) or 5.6 (Fig. 35) for different time intervals, in the presence of various amounts of F⁻, the different calcium dissolution curves showed two rates. In the first 20 minutes, the solubilized calcium amounts increased very quickly. This initial fast rate is followed by a slower phase with a nearly horizontal slope of the dissolution curves. The synthetic apatites were more fluoride-reactive than the enamel crystals. The familiar solubility reduction induced by F⁻ was observed.

Electron microscopic observations of dissolving synthetic hydroxyapatites after 15 minutes or 1 hour of buffer treatment at pH 5.6 confirmed that the initial fast dissolution step is coincident with the occurrence

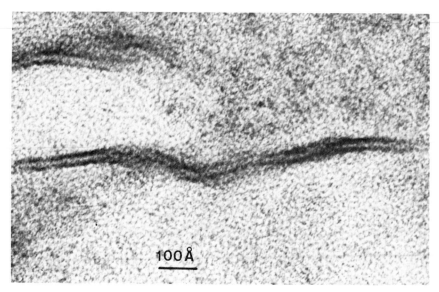

Fig. 33.   Longitudinal section of bone crystals during bone resorption.   Through destruction of the central part, the bone monocrystal is split into two thin plates.   x 1,050,000

of central core lesions, whereas the second slow phase corresponded to the appearance of lateral side crystal dissolutions.

VII.   DIFFRACTION STUDIES OF NORMAL AND CARIOUS HUMAN
       ENAMEL APATITES

That dissolution of the apatite crystals follows highly systematic crystallographic mechanisms was further confirmed by a single crystal electron diffraction investigation where the unit cell parameters of normal human apatite crystals were compared with those of adjacent carious crystals (7).   Carious enamel apatite crystals were assumed to be those presenting a central core lesion (Fig. 21,22,23).

Whereas no difference was noted in the values of the c axes, a highly significant difference ($p < 0.0001$) was observed in the $a = b$ parameters with an increase

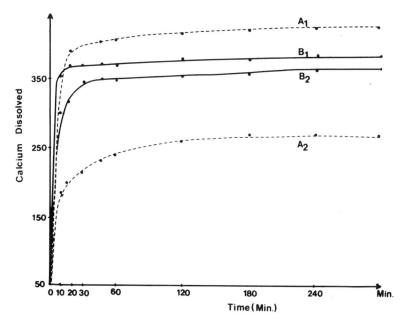

Fig. 34.   Dissolution curves at pH 3.7 of synthe-tic hydroxyapatite (A) and enamel powder (B) in the presence of 0 ppm (A_1 and B_1) and 2 ppm (A_2 and B_2) of F^-.   The amounts of dissolved calcium are expressed in micromoles/liter and the times in minutes.

of 0.05 Å in the carious crystals (Table II). Notably a similar increase of $\underline{a} = \underline{b}$ axes was found by Arends (27) by x-ray diffraction of bovine enamel apatite treated with acid.

This increase in $\underline{a} = \underline{b}$ parameters can have several explanations. It can be related to ionic substitutions, to a looser packing of the crystalline ions or to a preferential ionic or molecular dissolution phenomenon. A preferential loss of carbonate ions has been reported by Hallsworth $\underline{et\ al}$. (28) in enamel caries whereas an enrichment in $\overline{HPO_4}^{2-}$ has been noted by Arends and Davidson (29). Unfortunately, a direct correlation between these chemical findings and the parameter vari-ations demonstrated by diffraction studies cannot be presently made.

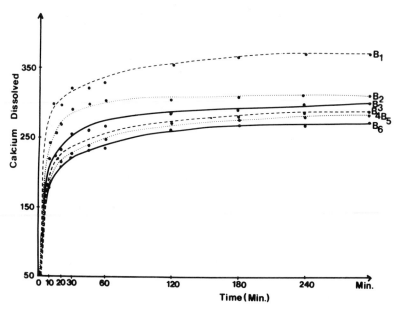

Fig. 35. Dissolution curves of enamel powder in a buffered solution at pH 5.6 in presence of 0 ppm (B_1), 2 ppm (B_2), 5 ppm (B_3), 7 ppm (B_4), 10 ppm (B_5) and 15 ppm (B_6) of F^-. The amounts of dissolved calcium are expressed in micromoles/liter and the times in minutes.

VIII.   CONCLUSION AND PERSPECTIVES

High resolution transmission electron microscopy reveals variations in shapes and sizes between human apatite crystals in enamel, dentin, and bone, although the basic lattice structures are similar. A clear trend from a hexagonal prismatic shape towards a true platelet configuration is apparent from enamel to dentin to bone crystals.

Similarity is found in the apparent mechanisms of crystal dissolution in dental caries and bone resorption. With increasing width to thickness ratios, one or several central holes develop rapidly along the <u>c</u>

TABLE II   COMPARISON OF THE UNIT CELL PARAMETERS a = b
AXIS OBTAINED BY SINGLE CRYSTAL ELECTRON DIFFRACTION
OF NORMAL AND CARIOUS ENAMEL (7) WITH SIMILAR VALUES
FOR BOVINE ENAMEL RECORDED BY ARENDS (27) AND
WITH X-RAY DIFFRACTION ON NORMAL AND
ACID-TREATED BOVINE ENAMEL

Normal Bovine Enamel (27)	Normal Human Enamel (7)
$c = 6.89 \pm 0.01$ Å $a = b = 9.44 \pm 0.01$ Å	$c = 6.88$ Å $a = b = 9.44 \pm 0.02$ Å
Acid Treated Bovine Enamel	Carious Human Enamel
$c = 6.88 \pm 0.01$ Å $a = b = 9.49 \pm 0.02$ Å	$c = 6.88$ Å $a = b = 9.49 \pm 0.02$ Å

axis. The extension of the central holes along (100)
planes in dentin and bone is very similar to the lateral
dissolution step occurring at slower velocity in enamel.
The crystallographic similarity between apatite dissolution during dental caries and bone resorption is not
surprising, if in both conditions organic acid demineralization be the prime etiological factor. Whereas in
caries, organic acids such as lactic, propionic, formic
and acetic acids have been suggested of significance, it
is assumed that citric acid is the agent responsible for
bone crystal dissolution (36).
     The various observations by high resolution elec-
tron microscopy and diffraction, as well as the chemical
kinetic evidence, indicate that biological apatite dissolution follows very precise rules and that the convergent use of advanced biophysical and biochemical
methods should lead to the identification of the preferential sites of dissolution. The specific blocking

or poisoning of these sites does not seem actually uto-
pian when considering the results obtained by Jonge
bloed (10) with EHDP diphosphonates on the acid dissolution of fluoroapatite single crystals.

IX.   REFERENCES

1. Nylen, M.U., and Omnell, K.A., in "Fifth Intern.
    Congr. for Electron Micr." (S.S. Breede Ed.),
    Vol. 2, p. QQ-4. Academic Press, New York, 1962.
2. Selvig, K.A., Calc. Tiss. Res. 6, 227 (1970).
3. Selvig, K.A., J. Ultrastr. Res. 41, 369 (1972).
4. Selvig, K.A., Z. Zellforsch. 137, 271 (1973).
5. Spector, M., J. of Microscopy 103, 55 (1975).
6. Hirai, G., and Fearnhead, R.W., in "Proc. 6th
    Conference on X-ray Optics and Microanalysis".
    p. 863. Univ. of Tokyo Press, 1972.
7. Voegel, J.C., and Frank, R.M., Caries Res. in
    press (1977).
8. Little, J.J., J. Roy. Micr. Soc. 78, 58 (1959).
9. Frazier, P.D., J. Ultrastr. Res. 22, 1 (1968).
10. Jongebloed, W.L., "An ultrastructural study of the
    caries process". Doctoraat Geneeskunde. Rijksuni-
    versiteit. Groningen, 1976.
11. Rönnholm, E., J. Ultrastr. Res. 6, 249 (1962).
12. Frank, R.M., Sognnaes R.F., and Kern, R., in
    "Calcification in Biological Systems" (R.F.
    Sognnaes, Ed.), p. 163. Am. Assoc. Adv. Science,
    Washington, 1960.
13. Nylen, M.U., J. Roy. Microsc. Soc. 83, 135 (1964).
14. Grove, C.A., Judd, G., and Ansell, G.S., J. Dent.
    Res. 51, 22 (1972).
15. Johansen, E., and Parks, H.F., J. Dent. Res. 39,
    714 (1960).
16. Johansen, E., J. Dent. Res. 6, 1007 (1964).
17. Selvig, K.A., and Halse, A., J. Ultrastr. Res. 40,
    527 (1972).
18. Simmelink, J.W., Nygaard, V.K., and Scott, D.B.,
    Archs Oral Biol. 19, 183 (1974).
19. Frank, R.M., and Voegel, J.C., in "Physico-chimie
    et cristallographie des apatites d'intérêt biolo-
    gique" (G. Montel Ed.) Colloques C.N.R.S. 230,

   p. 369, C.N.R.S. Paris, 1975.
20. Steve-Bocciarelli, D., Calc. Tiss. Res. 5, 261
   (1970).
21. Voegel, J.C., and Frank, R.M., J. Biol. Buccale
   2, 39 (1974).
22. Voegel, J.C., and Frank, R.M., Calc. Tiss. Res.
   in press (1977).
23. Arends, J., Royce, B.S.H., Welch, D.O., and
   Smoluchowski, R., in "Int. Symp. on the Structural
   Properties of Hydroxyapatite" (R.A. Young Ed.)
   Benjamin. New York, 1971.
24. Bonucci, E., Calc. Tiss. Res. 16, 13 (1974).
25. Michel, C., Parodontologie. 23, 191 (1969).
26. Garnier, P., Voegel, J.C., and Frank, R.M., J.
   Biol. Buccale 4, 323 (1976).
27. Arends, J., J. Dent. Res. 55, D 151 (abstract).
   (1976).
28. Hallsworth, A.S., Weatherell, J.A., and Robinson,
   C., Caries Res. 7, 345 (1973).
29. Arends, J., and Davidson, C.L., Calc. Tiss. Res.
   18, 65 (1975).
30. Irving, J.T., "Calcium and phosphorus metabolism",
   p. 69. Academic Press. New York. 1973.

DEGRADATIVE PROCESSES OF BONE

Paul Goldhaber, Luka Rabadjija and George Szabó

Harvard School of Dental Medicine

I. INTRODUCTION

Following pioneering studies of bone resorption by Fell and Mellanby (1), using vitamin A acetate or alcohol, and by Gaillard (2), using parathyroid tissue or extract, there has been great interest in the study of bone in vitro. However, so many factors have been discovered to influence bone resorption in tissue culture that at times it has appeared that nearly any compound may affect it. In addition, because different tissue culture systems have been employed in different laboratories, it has not always been possible to reproduce results between laboratories. Of course, caution must be exercised in extrapolating in vitro data and conclusions drawn from them to the in vivo behavior of bone. Nevertheless, the tissue culture approach to bone physiology has provided a powerful tool and thus insight to the complexity of bone as a tissue. It is likely to continue to provide a rational approach to understanding the etiology and treatment of bone diseases. This paper will briefly review some of the factors that influence bone resorption in tissue culture and will describe recent studies in our laboratory suggesting that macrophages found in bone play a more significant role in its resorption than formerly considered.

It is clear that any mechanism proposed for the destruction of bone must provide first for the removal

of bone mineral and second, for the removal of its
collagen matrix. It is rather tempting, indeed, when
examining a histological section of several osteoclasts
in Howship's lacunae of bone, to hypothesize that the
osteoclasts may simply secrete an organic acid (either
lactic or citric) to dissolve the bone mineral and may
also secrete collagenase and perhaps other proteases to
digest the collagen. The development in our laboratory
of a tissue culture system using parathyroid extract
(PTE) as a standard for stimulating bone resorption and
comparing the influence of various factors on the
enhancement or inhibition of this process has permitted
exploration of the mechanisms involved (3).

II. MINERAL PHASE DISSOLUTION: CITRIC ACID VERSUS LACTIC ACID

In a collaborative study with Martin et al. (4),
it was shown that citrate levels in the media of PTE-
treated calvaria of neonatal mice were increased and
paralleled calcium and phosphorus alterations throughout the experimental bone resorption process. The
accumulation of citrate was explained in part by a
decreased oxidation of citrate brought about by the
PTE. Lactate production, on the other hand, was
enhanced in the resorbing cultures only during the
first four days of incubation, reflecting an increased
utilization of glucose. Similar results had been
obtained in an earlier study (5) in which parathyroid
extract was not involved, but in which bone resorption
(in a chicken embryo extract-containing medium) was
initiated merely by incubating stationary cultures
under elevated oxygen tension (95% O_2/5% CO_2 by comparison with 95% air/5% CO_2). Eight to 19 times as much
citric acid was released into the medium as was present
in the fresh calvaria at the start of the experiment,
indicating that the citric acid was produced metabolic-
ally in response to the high oxygen tension in the
resorbing cultures. Notably, the nonresorbing control
cultures exposed to air had significantly more lactic
acid accumulation in the medium than those exposed to
elevated oxygen levels. Indirect evidence also

suggested citric acid as the most important organic
acid involved in removal of bone mineral: 1) Vastly
larger numbers of mitochondria, presumably involved in
Krebs Cycle metabolism are found in osteoclasts as
compared to other bone cells (6); 2) Our bone resorption system is oxygen-dependent (7); 3) Bone resorption
is inhibited by non-toxic concentrations of 2,4-dinitro-
phenol (an uncoupler of oxidative phosphorylation) or
malonate (a powerful inhibitor of the Krebs Cycle
enzyme succinic dehydrogenase) (7).

III. MOUSE BONE COLLAGENASE

The discovery by Gross and Lapiere (8) of an
animal collagenase isolated from resorbing tadpole
tails led our laboratory to search for a similar enzyme
in resorbing bone cultures. Collagenolytic activity
from actively resorbing bone was detected in the media
after culturing the tissue with purified, reconstituted
H^3-hydroxyproline- and H^3-proline-labeled collagen
fibrils (9). In addition, by using H^3-proline-labeled
calvaria other data established that during active bone
resorption bone collagen is partially degraded to
peptides similar in size and sequence to those produced
by the action of bacterial collagenase on collagen
(10).
In 1969 a collagenase, similar to tissue collagen-
ases from other sources, was isolated and partially
purified from the culture media of actively resorbing
mouse tibiae (11). Further purification and characterization of the enzyme indicated that it had a molecular
weight of about 41,000 daltons and was similar to
collagenase fraction A obtained from the synovial fluid
of rheumatoid arthritis patients (12,13). However, in
contrast to the latter enzyme, mouse bone collagenase
activity was readily inhibited by serum.

IV. PHAGOCYTOSIS BY OSTEOCLASTS AND MACROPHAGES

Question has frequently been raised as to the phagocytic potential of the various bone cells, particularly osteoclasts (14). Studies in resorbing bone cultures exposed to dilute India ink (15) have demonstrated the presence of a moderate number of ink particles in some (but not all) osteoclasts (Fig. 1).

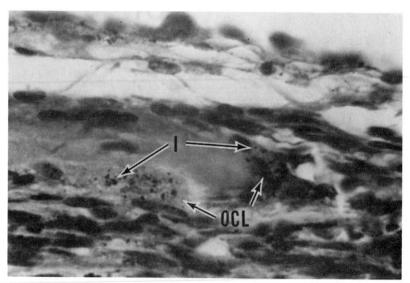

Fig. 1. Resorbing bone of mouse calvarium. Tissue was briefly immersed in 5-fold diluted Higgins India ink prior to culture for 7 days in presence of 0.5 U PTE/ml. Section stained by hematoxylin and eosin. Note resorption of bone (B) by two osteoclasts (OCL). Both cells show India ink particles in the cytoplasm. Primary magnification about 630X.

Other cells, presumably macrophages and fibroblasts,
were more heavily labeled. This observation is in
partial agreement with the old findings of McLean and
Bloom (16) who found macrophages "packed" with bone
mineral but osteoclasts relatively free of such material 12 hours after stimulating bone resorption in a
puppy with an injection of a toxic dose of parathyroid
extract.

V. OXYGEN, HEPARIN AND COLLAGENASE INTERACTIONS

As pointed out above, the oxygen tension in our
bone cultures plays a crucial role in facilitating or
inhibiting the action of any bone resorbing factor.
Heparin is almost as significant as oxygen. In our
tissue culture system heparin markedly enhances the
effectiveness of all bone resorption-stimulating factors without having appreciable intrinsic bone resorption activity (17). This property is shared by a
number of other highly charged polyanions such as the
synthetic compound Treburon (the sodium salt of sulfated polygalacturonic acid methyl-ester methyl glyco-
side) or dextran sulfate.
As for the mechanism of action of heparin as a
bone resorption stimulating co-factor, evidence
suggested that it might be linked to collagenase
activity. Not only did heparin increase the amount of
collagenase release into incubation media by bone
explants, it also directly enhanced the enzymatic
activity of mouse bone collagenase on solid collagen
substrate (18). Further studies of the interaction
between heparin and mouse bone collagenase (19) revealed that the enzyme could be tightly bound to a
heparin-substituted gel at low ionic strength and
eluted at high ionic strength. This technique incident-
ally provided a method for further purification of
mouse bone collagenase with high yield and for isolation and measurement of its activity in our tissue
culture media, despite the presence of serum inhibitors
such as α-2-macroglobulin (20). Using this technique,
it has been possible to correlate the extent of bone
resorption in living bone cultures with the amount of

collagenase activity found in the culture media (21).
These same investigators are conducting fluorescent
antibody studies to localize mouse bone collagenase
within the resorbing bone tissue.

Nagai et al. (22) have proposed that collagenase
activity in vivo is regulated by α-2-macroglobulin and
that the enzyme escapes this inhibition after secretion
from the enzyme-forming cells so long as it is bound to
collagen fibers. Possibly another way of avoiding
inhibition is for the enzyme to be transported to the
collagen fibers in a latent form to be subsequently
activated enzymatically.

VI.   PROSTAGLANDINS AND BONE RESORPTION

In recent years one of the most exciting observations in bone research has been the finding that prostaglandins, particularly of the E series, are potent
stimulators of bone resorption in culture (23). This
work has led to the implication of prostaglandins in
the bone loss associated with periodontal disease (24,
25), the hypercalcemia of malignancy (26,27), dental
cysts (28,29), and rheumatoid arthritis (30). With
regard to rheumatoid arthritis, it should be noted that
rheumatoid synovia produce approximately 10 times
more prostaglandin E_2 in culture than normal synovia.
Presumably the pharmacologic effects of nonsteroid
anti-inflammatory drugs such as indomethacin or aspirin
is due to their ability to inhibit prostaglandin
synthetase, thereby blocking the continuous production
of prostaglandin E_2 and the subsequent bone destruction
associated with rheumatoid arthritis (30).

VII.   ENDOTOXIN AND BONE RESORPTION

The finding that highly purified bacterial endotoxins can stimulate significant bone resorption (31)
forged a strong link between microorganisms, their
products, and their possible involvement in periodontal
disease. Further work established that it is the
lipid A portion of the lipopolysaccharide molecule that

is responsible for the bone resorption stimulating
activity (32). Others showed that guinea pig peritoneal exudate macrophages produce collagenase when
exposed to bacterial lipopolysaccharides or to their
lipid A fraction (33). Products from antigen- or
mitogen-stimulated lymphocytes also activate macrophages to synthesize collagenase (34). Mouse peritoneal macrophages can be stimulated by thioglycollate,
latex particles, or dextran sulfate to secrete a
specific collagenase in culture (35). This finding
with respect to dextran sulfate is consistent with its
heparin-like properties and the ability of heparin to
enhance collagenase release in bone cultures (18). In
view of the abnormal connective tissue resorption
associated with rheumatoid arthritis, it is not surprising that synovial explants in culture produce collagenase (36), probably originating from the macrophages in
the tissue. However, Krane (37) cautions that while
the enzyme can attack demineralized bone collagen it
has no effect on mineralized bone collagen. Therefore,
a mechanism is required for the removal of the inorganic
component of bone before collagenase action.

VIII. OSTEOCLAST ACTIVATING FACTOR AND BONE RESORPTION

        Horton and collaborators also linked immunologi-
cal phenomena to bone resorption by demonstrating that
supernatant fluids from cultures of human leukocytes
activated either by phytohemagglutinin or by antigenic
materials in dental plaque contain a mediator, called
"osteoclast activating factor" (OAF), which causes
osteoclasts to resorb bone in culture (38). Further
studies by this group demonstrated that OAF production
requires cellular interaction between macrophages and
lymphocytes; supernatants of cultured macrophages are
unable to stimulate lymphocytes to produce the mediator.
These investigators favor the notion that lymphocytes
rather than macrophages produce OAF, because irradiation of a mixed population of macrophages and lymphocytes markedly decreased OAF production and because
lymphocytes are much more radiosensitive than macrophages (39). Since indomethacin does not inhibit the

bone resorption stimulatory effects of OAF, it was
concluded that prostaglandin synthesis is not involved
in the function of this mediator (40).

IX. GOLD SALTS AND RHEUMATOID ARTHRITIS

Both indomethacin and aspirin inhibit gingival
fragment media-stimulated bone resorption in tissue
culture (24). Therefore, it was deemed worthwhile to
test the effect on bone resorption of other compounds
used in the treatment of rheumatoid arthritis. We were
especially interested in the use of gold salts because
we felt they might be useful as markers. This form of
treatment (chrysotherapy) has long been used, appears
to be gaining in popularity, and is reported to be
helpful (41). A substantial literature exists testify-
ing to the efficacy of chrysotherapy but only limited
attempts to explain the mechanism of action of gold
salts have been made. Persellin and Ziff (42) demonstrated that sodium aurothiomalate (SATM; Myochrysine,
Merck, Sharp and Dohme, West Point, PA.) inhibits two
lysosomal enzymes from guinea pig macrophages, acid
phosphatase and β-glucuronidase. This was demonstrated
both with lysed and intact cells. Some ultrastructural
studies attempting to localize the gold morphologically
to the macrophage lysosomes have had equivocal results
(43). However, Oryschak and Ghadially (44) showed that
injections of SATM into rabbit knee joints produced
characteristic chondrocyte lysosomes (called aurosomes)
containing gold inclusions, as determined by electron
probe x-ray analysis. Similarly, studies on human
synovial tissues injected with radioactive colloidal
gold showed that the tracer was distributed in the
lysosomes of synovial macrophages shortly after administration and remained there for years (45).

X.   EFFECT OF GOLD SALTS ON BONE RESORPTION IN TISSUE
     CULTURE

     SATM was tested in our bone resorption culture
system (46), utilizing the calvaria of 5-day old mice,
in an attempt to determine whether this gold salt could
inhibit PTE-stimulated resorption. The degree of
resorption was determined both morphologically, by a
modification of the method of Susi et al. (47), and
chemically, using a Corning Calcium Analyzer (Model 940)
to measure the total calcium released from the resorbing bones during the entire culture period. The
results of a typical experiment with SATM and PTE are
shown in Figure 2. PTE (0.1 U/ml) stimulated significantly more resorption than found in the control. SATM
exhibited a dose-dependent response in its inhibition
of the PTE-induced resorption. Although 10 µg SATM/ml
had no effect, 50 µg/ml reduced the resorption to less
than half of the PTE control level, and 100 µg/ml
completely abolished the PTE effect. The calcium
release data closely paralleled the morphological
scoring data. Control experiments were performed,
testing the effect of thiomalic acid and benzyl alcohol
(present as a preservative in the Myochrysine), both
separately and in various combinations, to determine
whether they could inhibit PTE-induced resorption.
They were without effect.
     In order to determine whether the inhibition of
PTE-stimulated resorption by SATM was merely due to a
generalized, non-specific toxic effect, this agent was
tested in our bone remodeling system (48). This in
vitro system normally demonstrates a moderate amount of
resorption (without the addition of PTE) and also
exhibits significant new osteoid formation during a
14-day culture period. Selective inhibition of resorption and the maintenance of new bone formation by a
compound tested in this system indicates that the test
compound specifically inhibits the bone resorption
process and is not merely killing the bone. Figure 3
shows the effect of SATM on bone resorption in the
remodeling system after 14 days. Although the resorption found in the control group was not as severe as
that caused by PTE (Fig. 2) it was still substantial.

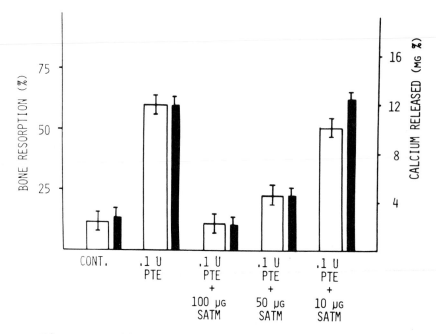

Fig. 2. Effect of various concentrations of sodium aurothiomalate (SATM) on parathyroid extract (PTE)-stimulated bone resorption in tissue culture. Open bars indicate mean bone resorption score ± SEM; solid bars indicate mean calcium release ± SEM. Note the marked enhancement of resorption by PTE over the basal control value as measured by morphological scoring and calcium release into the medium. Also note the inhibition of PTE resorption (dose-response) by SATM at 100 µg/ml and 50 µg/ml.

In addition, the resorption-inhibition effect of SATM was clearly evident at concentrations of 50 µg/ml or 100 µg/ml. Again, the calcium release values paralleled the morphological scores except at 10 µg/ml, at which the calcium release was somewhat depressed compared to the untreated control.

Microscopic examination of histological sections from this experiment revealed good new osteoid formation after 14 days both in the untreated control group and in the SATM-treated groups, thereby suggesting that SATM was acting specifically on the resorption process.

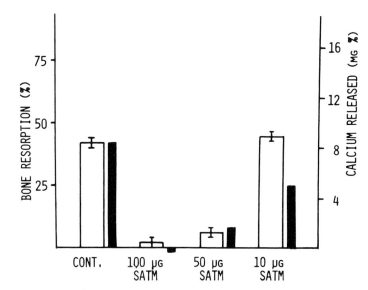

Fig. 3. Effect of various concentrations of sodium aurothiomalate (SATM) on bone resorption in the bone remodeling system after 14 days culture. Open bars indicate mean bone resorption score ± SEM; solid bars indicate single pooled values of calcium release per group obtained by analysis of tissue culture media. Note the inhibition of resorption (dose-response) due to SATM at 100 μg/ml and 50 μg/ml.

However, the most striking observation (particularly in the 100 μg/ml group) was the conspicuous number of macrophages scattered throughout the section, many of them filled with yellowish-brown particles, inferred to be ingested gold. In order to determine whether the ingestion of gold had occurred during an early phase, when bone resorption would ordinarily be maximal, or whether the uptake of gold by macrophages was limited to a later phase of the culture period, another experiment was performed with the remodeling system and terminated at 4 days. The results were almost identical to the 14-day experiment: SATM at 50 μg/ml or 100 μg/ml inhibited resorption and numerous macrophages conspicuously contained yellowish-brown particles.

Histological examination of bone cultures exposed to 100 µg SATM/ml indicated diminished cellularity and fewer osteoclasts than untreated controls. Preliminary ultrastructural studies of 4-day and 14-day bone cultures exposed to 50 µg SATM/ml or 100 µg SATM/ml revealed the presence of cells containing numerous electron-dense bodies in secondary lysosomes (auro-somes?) and vesicles (Fig. 4). These findings are similar to those of Oryschak and Ghadially (44) and of Yarom et al (45) and support the concept that gold salts concentrate in secondary lysosomes or vacuoles of macrophages.

In view of the report of Persellin and Ziff (42) that SATM inhibits the lysosomal acid phosphatase of guinea pig macrophages, representative histological sections were stained for acid phosphatase according to the histochemical method of Barka and Anderson (49). The control resorbing bones cultured for 4 days exhib-ited intensive staining of the osteoclasts and, in some areas, of the margin of the resorbed or resorbing bone. On the other hand, sections from bones treated with 100 µg SATM/ml were almost devoid of acid phosphatase stain, except for a rare Howship's lacuna margin and some small cells nearby. At 50 µg SATM/ml, acid phos-phatase staining was almost as intense as for the untreated control.

XI. AN ALTERNATIVE VIEW OF BONE RESORPTION

The results of these recent tissue culture studies establish that non-toxic concentrations of SATM inhibit bone resorption. Although the effective concentrations are higher than those normally reported in sera of patients treated with gold salts, it is obvious that the amounts of gold found in phagocytic cells far exceed reported serum levels (50).

It is tempting to explain the mechanism of action of SATM as an interference with lysosomal enzymes. Vaes (51) hypothesized that the lysosomes contain all the enzymes required to solubilize or digest the organic components of bone matrix. However, as an alternative proposal, he suggested that there may be

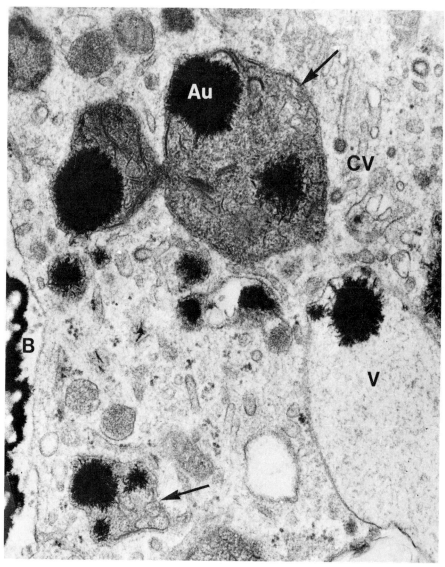

Fig. 4. Ultrastructure of cell adjacent to bone
(B) cultured 4 days in the presence of 50 µg of SATM/ml
of medium. Electron-dense bristling clumps (AU) and
spicules consistent with the presence of gold are found
in secondary lysosomes (arrows) and in vacuoles (V).
Note coated vesicles (CV). Ito-Karnovsky fixative;
stained with uranyl acetate and lead citrate, X 45,000.

two mechanisms to control bone resorption, one involving the exocytosis of the lysosomal enzymes and the
other involving the independent secretion of a collagenase from a nonlysosomal source. Jessop et al. (52),
using a "skin-window" technique, reported that the
phagocytic activity of macrophages and neutral poly-
morphs is elevated in rheumatoid arthritis and
progressively suppressed during treatment with gold
salts. They ascribed this finding to a "progressive
saturating effect on these cells".

The decreased general cellularity and especially
the decreased numbers of osteoclasts in bone cultures
exposed to 100 µg SATM/ml can be related to the results
of Viken and Lamvik (53), who exposed human monocytic
phagocytes to SATM at various stages of differentiation
in vitro. These workers observed a marked depression
of differentiation of monocytes to macrophages upon
exposing the cells continuously to SATM from 90 minutes
to 8 days in culture. If monocytes or macrophages do,
indeed, differentiate further by fusion to become
osteoclasts (54), it is conceivable that high concentrations of SATM in precursor cells within the bone
cultures could block osteoclast formation and function,
explaining its inhibition of bone resorption.

Recently, Lipsky and Ziff (55) demonstrated that
SATM inhibits in vitro antigen- and mitogen-triggered
human lymphocyte DNA synthesis. SATM had to be present
at the initiation of the cultures and the inhibition
could be reversed by supplementation of the cultures
with purified monocytes. They suggested that gold
prevents T-lymphocyte activation by interfering with
the ability of monocytes to serve as effective accessory
cells in the initiation of the lymphoproliferative
response. These investigators concluded that the
efficacy of gold compounds in reducing the immunologically mediated chronic inflammation of rheumatoid
arthritis is due to their interference with mononuclear
phagocytes in the induction of both cellular and humoral
immune responses. Could it be that a similar interaction between macrophages and lymphocytes or between
macrophages and osteoclasts (or among all three cell
types) is required for active bone resorption to occur?
Experiments carried out some years ago in our laboratory

seem to rule out an absolute requirement for lymphocytes. We attempted to determine whether various doses
of x-irradiation in vitro would have a differential
effect on subsequent bone resorption and formation in
cultured mouse calvaria. Although osteoid formation
was entirely inhibited after about 1,000 rads, bone
resorption was still very active after doses as high as
10,000 rads (56). Clearly, any lymphocytes associated
with the calvaria at the time of explantation would have
been killed with some of the higher doses used. These
data therefore indicate that the in vitro bone resorption observed was not dependent on the release of
mediator from viable lymphocytes. On the other hand,
macrophages could survive such high doses of x-irradiation and were most prominent in histological sections of
these resorbing bones.

The possible interaction or "teamwork" between
macrophages and osteoclasts as a requirement for bone
resorption is thus a more appealing hypothesis. The
osteoclasts could provide the organic acid (citric?)
to demineralize the bone while the activated macrophages
could provide the collagenase to help destroy the
collagen matrix. These two cell types, working in
concert, could affect bone resorption.

An alternate hypothesis would hold that either the
macrophage or the osteoclast could resorb bone
independently. Perhaps, however, too much significance
has been attributed to osteoclasts in bone resorption
to the exclusion of other cells. Although Goldhaber (7)
clearly demonstrated by means of time-lapse microcine-
matography the active participation of osteoclasts in
bone resorption (particularly in the formation of
Howship's lacunae) he also stated, "It should be noted,
however, that although macrophages are much smaller than
osteoclasts, because of their better contrast at this
magnification, it has been possible on occasion to
distinguish these very motile cells attacking bone
spicules and apparently participating directly in the
resorptive process. These observations support the
histological findings of McLean and Bloom, but in
addition ascribe to macrophages bone-resorbing abilities
beyond mere scavenging of debris."

XII. ACKNOWLEDGEMENTS

The authors wish to express their deep apprecia-
tion to Joan Jennings, Lorraine Stevens, and Evelyn
Flynn for their outstanding technical assistance.
This work was supported in part by U.S. Public
Health Service Grants Nos. DE-02849 and DE-01766,
National Institute of Dental Research.

XIII. REFERENCES

1. Fell, H.B., and Mellanby, E., J. Physiol. (London)
    116, 320 (1952).
2. Gaillard, P.J., Exp. Cell Res. Suppl. 3, 154
    (1955).
3. Goldhaber, P., in "The Parathyroid Glands" (P.J.
    Gaillard, R.V. Talmage, and A.M. Budy, Eds.),
    p. 153. The University of Chicago Press, Chicago,
    1965.
4. Martin, G.R., Mecca, C.E., Schiffmann, E., and
    Goldhaber, P., in "The Parathyroid Glands" (P.J.
    Gaillard, R.V. Talmage, and A.M. Budy, Eds.),
    p. 261. The University of Chicago Press, Chicago,
    1965.
5. Kenny, A.D., Draskoczy, P.R., and Goldhaber, P.,
    Am. J. Physiol. 197, 502 (1959).
6. Cameron, D.A., in "The Biochemistry and Physiology
    of Bone" (G.H. Bourne, Ed.), Second Edition, Vol.
    1, Structure. p.191. Academic Press, New York,
    1972.
7. Goldhaber, P., in "The Parathyroids" (R.O. Greep,
    and R.V. Talmage, Eds.), p. 243. Charles C.
    Thomas, Springfield, IL, 1961.
8. Gross, J., and Lapiere, C.M., Proc. Natl. Acad.
    Sci. U.S.A. 48, 1014 (1962).
9. Kaufman, E.J., Glimcher, M.J., Mechanic, G.L., and
    Goldhaber, P., Proc. Soc. Exp. Biol. Med. 120,
    632 (1965).
10. Stern, B.D., Glimcher, M.J., Mechanic, G.L., and
    Goldhaber, P., Proc. Soc. Exp. Biol. Med. 119,
    577 (1965).

11. Shimizu, M., Glimcher, M.J., Travis, D., and
     Goldhaber, P., Proc. Soc. Exp. Biol. Med. 130,
     1175 (1969).
12. Sakamoto, S., Goldhaber, P., and Glimcher, M.J.,
     Calcif. Tissue Res. 10, 142 (1972).
13. Harris, E.D., Jr., DiBona, D.R., and Krane, S.M.,
     J. Clin. Invest. 48, 2104 (1969).
14. McLean, F.C., and Urist, M.R., "Bone - An
     Introduction to the Physiology of Skeletal Tissue",
     Second Edition, p. 93. The University of Chicago
     Press, Chicago, 1961.
15. Goldhaber, P., in "In Vitro: Differentiation and
     Defense Mechanisms in Lower Organisms." A
     Symposium of the Tissue Culture Association.
     Philadelphia, June 5-7, 1967" (M.M. Sigel, Ed.),
     Vol. 3, p. 178. Williams and Wilkins Co.,
     Baltimore, MD., 1968.
16. McLean, F.C., and Bloom, W., Arch. Pathol. 32, 315
     (1941).
17. Goldhaber, P., Science 147, 407 (1965).
18. Sakamoto, S., Goldhaber, P., and Glimcher, M.J.,
     Calcif. Tissue Res. 12, 247 (1973).
19. Sakamoto, S., Sakamoto, M., Goldhaber, P., and
     Glimcher, M.J., Biochim. Biophys. Acta 385, 41
     (1975).
20. Sakamoto, S., Goldhaber, P., and Glimcher, M.J.,
     Calcif. Tissue Res. 10, 280 (1972).
21. Sakamoto, S., Sakamoto, M., Goldhaber, P., and
     Glimcher, M., Biochem. Biophys. Res. Commun. 63,
     172 (1975).
22. Nagai, Y., Hori, H., Kawamoto, T., and Komiya, M.,
     in "Dynamics of Connective Tissue Macromolecules"
     (P.M.C. Burleigh, and A.R. Poole, Eds.), p. 171.
     North-Holland Publishing Co., Amsterdam, 1975.
23. Klein, D.C., and Raisz, L.G., Endocrinology 86,
     1436 (1970).
24. Goldhaber, P., Rabadjija, L., Beyer, W.R., and
     Kornhauser, A., J. Am. Dent. Assoc. 87 (Special
     Issue), 1027 (1973).
25. Goodson, J.M., Dewhirst, F.E., and Brunetti, A.,
     Prostaglandins 6, 81 (1974).
26. Tashjian, A.H., Jr., Voelkel, E.F., Levine, L.,
     and Goldhaber, P., J. Exp. Med. 136, 1329 (1972).

27. Tashjian, A.H., Jr., Voelkel, E.F., Goldhaber, P.,
     and Levine, L., Prostaglandins 3, 515 (1973).
28. Harris, M., and Goldhaber, P., Br. J. Oral Surg.
     10, 334 (1973).
29. Harris, M., Jenkins, M.V., Bennett, A., and Wills,
     M.R., Nature (London) 245, 213 (1973).
30. Robinson, D.R., Tashjian, A.H., Jr., and Levine,
     L., J. Clin. Invest. 56, 1181 (1975).
31. Hausmann, E., Raisz, L.G., and Miller, W.A.,
     Science 168, 862 (1970).
32. Hausmann, E., Lüderitz, O., Knox, K., and Weinfeld,
     N., J. Dent. Res. 54, Special Issue B, B94 (1975).
33. Wahl, L.M., Wahl, S.M., Mergenhagen, S.E., and
     Martin, G.R., Proc. Natl. Acad. Sci. U.S.A. 71,
     3598 (1974).
34. Wahl, L.M., Wahl, S.M., Mergenhagen, S.E., and
     Martin, G.R., Science 187, 261 (1975).
35. Werb, Z., and Gordon, S., J. Exp. Med. 142, 346
     (1975).
36. Harris, E.D., Jr., and Krane, S.M., Arthritis
     Rheum. 14, 669 (1971).
37. Krane, S.M., in "Dynamics of Connective Tissue
     Macromolecules" (P.M.C. Burleigh, and A.R. Poole,
     Eds.), p. 309, North-Holland Publishing Co.,
     Amsterdam, 1975.
38. Horton, J.E., Raisz, L.G., Simmons, H.A.,
     Oppenheim, J.J., and Mergenhagen, S.E., Science
     177, 793 (1972).
39. Horton, J.E., Oppenheim, J.J., Mergenhagen, S.E.,
     and Raisz, L.G., J. Immunol. 113, 1278 (1974).
40. Mergenhagen, S.E., Wahl, S.M., Wahl, L.M., Horton,
     J.E., and Raisz, L.G., Ann. N.Y. Acad. Sci. 256,
     132 (1975).
41. Article, Br. Med. J. 2 (5964), 156 (1975).
42. Persellin, R.H., and Ziff, M., Arthritis Rheum. 9,
     57 (1966).
43. Norton, W.L., Lewis, D.C., and Ziff, M., Arthritis
     Rheum. 11, 436 (1968).
44. Oryschak, A.F., and Ghadially, F.N., Virchows Arch.
     B Zellpathol. 17, 159 (1974).
45. Yarom, R., Hall, Th. A., Stein, H., Robin, G.C.,
     and Makin, M., Virchows Arch. Abt. B Zellpathol.
     15, 11 (1973).

46. Goldhaber, P., J. Dent. Res. 50, 278 (1971).
47. Susi, F.R., Goldhaber, P., and Jennings, J.M.,
    Am. J. Physiol. 211, 959 (1966).
48. Goldhaber, P., J. Dent. Res. 45, 490 (1966).
49. Barka, T., and Anderson, P.J., J. Histochem.
    Cytochem. 10, 741 (1962).
50. Jessop, J.D., and Johns, R.G.S., Ann. Rheum. Dis.
    32, 228 (1973).
51. Vaes, G., in "Lysosomes in Biology and Pathology"
    (J.T. Dingle, and H.B. Fell, Eds.), Frontiers of
    Biology, 14A. Vol. 1, p. 217. North-Holland
    Publishing Co., Amsterdam, 1969.
52. Jessop, J.D., Vernon-Roberts, B., and Harris, J.,
    Ann. Rheum. Dis. 32, 294 (1973).
53. Viken, K.E., and Lamvik, J.O., Acta Pathol.
    Microbiol. Scand. Sect. C, 84, 419 (1976).
54. Hancox, N.M., in "The Biochemistry and Physiology
    of Bone" (G.H. Bourne, Ed.), Second Edition, Vol.
    1, Structure. p. 45. Academic Press, New York,
    1972.
55. Lipsky, P.E., and Ziff, M., J. Clin. Invest. 59,
    455 (1977).
56. Goldhaber, P., and Little, J.B., J. Cell Biol. 27,
    35A (1965).